Ecotaxes on Polluting Inputs and Outputs

Published by
Academic Foundation
in association with the
Madras School of Economics, Chennai

Ecotaxes

on Polluting Inputs and Outputs

Raja J. Chelliah
Paul P. Appasamy
U. Sankar
Rita Pandey

ACADEMIC FOUNDATION
NEW DELHI

First published in 2007
by

ACADEMIC FOUNDATION
4772-73 / 23 Bharat Ram Road, (23 Ansari Road),
Darya Ganj, New Delhi - 110 002 (India).
Phones : 23245001 / 02 / 03 / 04.
Fax : +91-11-23245005.
E-mail : academic@vsnl.com
www. academicfoundation.com

in association with:

Madras School of Economics, Chennai
www.mse.ac.in

Cataloging in Publication Data--DK
Courtesy: D.K. Agencies (P) Ltd. <docinfo@dkagencies.com>

Ecotaxes on polluting inputs and outputs /
editors, Raja J. Chelliah ... [et al.].
p. cm.
Includes bibliographical references (p.)
ISBN-13: 9788171885848
ISBN-10: 8171885845

1. Environmental impact charges--India. 2. Pollutants--Economic aspects--India. 3. Environmental economics--India. I. Chelliah, Raja J. (Raja Jesudoss)

DDC 336.278 363 179 095 4 22

Typeset by Italics India, New Delhi.
Printed and bound in India.

Contents

...CONTD. ...

List of Figures, Tables and Annexures

FIGURES

TABLES

ANNEXURES

सत्यमेव जयते

डॉ. सी. रंगराजन
Dr. C. RANGARAJAN

अध्यक्ष
प्रधानमंत्री की आर्थिक सलाहकार परिषद्
विज्ञान भवन सौंध 'ई' हाल
मौलाना आज़ाद रोड
नई दिल्ली–110 011

CHAIRMAN
Economic Advisory Council to the Prime Minister
Vigyan Bhavan Annexe, 'E' Hall
Maulana Azad Road
New Delhi-110 011

Foreword

The National Environmental Policy (2006) has recommended the use of economic instruments to supplement environmental regulations. The Ministry of Environment and Forest, Government of India commissioned a study by the Madras School of Economics and the National Institute of Public Finance and Policy on levying eco-taxes on polluting inputs and outputs. This pioneering study coordinated by Prof. Raja J. Chelliah provides proposals for eco-taxes in various sectors such as coal, automobiles, phosphate based detergents, chlorine in paper, pulp and viscose rayon industries, pesticides, fertilisers, lead acid batteries and plastics. The major advantage of the proposals is that they utilise the existing tax machinery for central taxes. Revenues from the taxes could be recycled to assist industries to comply with environmental requirements. Tax exemptions for green substitutes will also encourage the widespread use of alternative inputs and outputs.

The book will be of use to policy makers, government agencies, industries, NGOs and other stakeholders interested in the field of environmental policy and management.

[C. Rangarajan]

Chairman, Madras School of Economics, Chennai
Chairman, National Institute of Public Finance and Policy, New Delhi

Acknowledgements

The authors are grateful to the Ministry of Environment and Forests, Government of India for funding the Centre of Excellence in Environmental Economics, Madras School of Economics to carry out this pioneering study on the use of ecotaxes on polluting inputs and outputs.

The following experts worked as Consultants in the Project on the areas assigned to them.

1. Mr. B.C. Rastogi—Expert on indirect taxes in all the sectors.
2. Ms. M. Karpagam—Literature Review on International Experience in Ecotaxes.
3. Dr. Vinish Kathuria—Automobiles.
4. Mr. S. Mukherjee—Study of emissions from diesel *versus* petrol engines.
5. Dr. S. Jayaraj—Pesticides.
6. Dr. Joseph Thomas—Pesticides.
7. Dr. S. Renganarayanan—Emissions from power plants.
8. Dr. R. Sethumadhavan—Emissions from power plants.

We thank them for their valuable contributions. Mr. Rastogi was associated with the project from the beginning and participated in many discussions. The Interim Report was also presented to a few government agencies and Industry Associations such as SIAM (automobiles); Paper manufacturers (Chlorine in paper and viscose rayon); Indian Soaps and Toiletries Makers' Association (Detergents); Fertiliser Association of India (Fertilisers) and Plastic Manufacturers Association (Plastics) for their comments and suggestions which have been incorporated in the report. The findings of the Report were discussed in a meeting convened by the Ministry of Environment and

Forests. We are grateful to the Secretary, Additional Secretary and Economic Adviser of the Ministry for their constructive comments and suggestions.

Raja J. Chelliah

U. Sankar

Paul P. Appasamy

Rita Pandey

About the Authors

Paul P. Appasamy .... *Member–Secretary, Centre of Excellence in Environmental Economics, Madras School of Economics, Chennai.*

Raja J. Chelliah .... *Chairman Emeritus, Madras School of Economics, Chennai and Professor of Eminence, National Institute of Public Finance & Policy, New Delhi.*

Rita Pandey .... *Professor, National Institute of Public Finance and Policy, New Delhi.*

U. Sankar .... *Honorary Professor, Madras School of Economics, Chennai.*

1

Introduction

Externality

Externality arises when consumption or production decisions of one economic unit enters into the utility or production function of another economic unit without any compensation. When a firm discharges its untreated effluents on water bodies or land, or its emissions into atmosphere, they result in harm/damage to society. The firm treats the environment as if it were a free good or an unpaid factor of production; it ignores the cost of degradation of the environment. Intertemporal negative externalities arise when exhaustible resources are depleted at rates faster than the social optimal rates and when renewable resources are harvested at rates faster than their regeneration rates.

Regulatory Approach

There are two broad approaches to solve this externality problem. The first approach is state regulation of activities causing negative externalities by prescription of source-specific standards, non-transferable quotas, restriction on location of activities, and ban on production of or/and use of certain inputs or processes. This type of regulation is called command and control (CAC). This type of regulation presupposes enormous information (which is seldom available) on the part of the regulator in prescribing standards/control measures. It restricts the choices available to polluters in their search for cost minimising solutions for pollution prevention and control. India's pollution control regime based on environmental legislations and rules emanating from the legislations may be characterised as 'prohibit and punish' regime.[1] In this regime the penalties for non-compliance with the regulations are unrelated

1. See Divan and Rosencranz (2001).

to the compliance costs. The enforcement takes the form of either full compliance or no compliance; the extent of violations (which determine the compliance cost) is not considered by the regulators and the courts.

There is a case for CAC regulation when the perceived harm from an activity is very high, as for example, in production or use of highly hazardous materials or when the assimilative capacity of a region in which a plant is to be located is reaching the threshold level or when an irreversible ecological damage would result because of the location of a project. In such cases the precautionary principle justifies CAC type of regulation.

Economic Approaches

Policy prescriptions for internalising the externalities are generally based on Pigouvian or Coasean approaches.

Pigouvian Approach

Pigou (1920) considered the case of a competitive firm producing a 'good' and a 'bad' (smoke). It is assumed that smoke emission per unit of output is constant. The smoke emitted into the atmosphere has a harmful effect on nearby residents and producers and they bear the costs in the form of higher health costs and production costs respectively. These costs are uncompensated by the polluting firm which treats the atmosphere as a sink, an unpaid factor of production. In the absence of any regulations, the equilibrium output of the industry is determined at the point where the private marginal cost is equal to the price. The socially optimal output is determined at the point where the marginal social cost is equal to the price. The socially optimal level of output of the polluting firm is smaller than the output in the absence of regulation. The Pigouvian policy prescription for the correction of the negative externality is imposition of a tax on the output equal to the difference between the marginal social cost and price at the socially optimal output. It can be shown that such a tax maximises aggregate welfare defined as sum of consumers' and producers' surpluses.[2] The imposition of a per

2. The same result can be obtained in a general equilibrium model using the Pareto Welfare criterion. According to Pareto social welfare increases if it is possible to improve the welfare of at least one individual without decreasing the welfare of others. For the general result, see Baumol and Oates (1987).

unit tax on the good gives a signal to the producers and consumers about the social harm resulting from the production of the good. The resulting rise in the price of the good gives an incentive to reduce the consumption of that good and therefore to reduce the pollution load. Pigou adds that, "sometimes, when the interrelations of the various private persons affected are highly complex, the government may find it necessary to exercise some means of authoritarian control" (p.194).

Coasean Approach

Coase (1960) argues that the problem which we face in dealing with actions which have harmful effects is not simply one of restraining those responsible for them. What has to be decided is whether the gain from preventing the harm is greater than the loss which would be suffered elsewhere as a result of stopping the action which produces the harm. He says it is necessary to know whether the damaging business is liable or not for the damage caused since without the establishment of this initial delimitation of rights there can be no market transactions to transfer and recombine them. But the ultimate result (which maximises the value of production) is independent of the legal position if the pricing system is assumed to work without cost. This proposition is known as 'Coase Theorem'.

Coase advocates a role for the state in defining and enforcing property rights for environmental resources and in mitigating transaction costs but rules out government intervention in the form of specifying standards or levying a tax to correct the externality. Creation of rights for emissions into atmosphere and setting up markets for exchange of the rights, and programmes such as green rating of industry and public disclosure of the extent of environmental compliance by polluting units are preferred under this approach.

Baumol and Oates-Second Best Approach

There are many conceptual and measurement problems in translating Pigouvian prescription into a policy instrument. Baumol and Oates (1987) note the following difficulties in applying the Pigouvian approach in practice. First, the very presence of externalities is likely to produce a number of local maxima and it seems impossible to choose any one with some degree of confidence. Second, the optimal tax level on an externality-generating activity is not equal to the marginal net damage generated initially, but rather to the damage it would cause if the level of the activity had been adjusted to its optimal level. In view of the enormous amount of information needed to estimate the marginal damage function, they propose an alternative approach which "consists of the use of a set of standards that serve as targets for environmental quality coupled with fiscal measures and other compensatory instruments used as a means to attain these standards" (p.159).

In fact most countries including India have adopted a second-best pollution control regime based on any one or a combination of the following approaches, namely,

(i) standards and regulation

(ii) standards and charges

(iii) standards and permits.

The superiority of economic instruments (EIs) (ii) and (iii) compared with CAC instruments (i) is that the former gives flexibility to the polluters in achieving the standards at least cost. Tietenberg (1985) and Anderson and Carlin (1999) and many others report cost savings from the adoption of EIs compared to the use of CAC instruments.

International and Domestic Support for Economic Instruments

The Organisation for Economic Cooperation and Development (OECD) adopted in 1972 the 'polluter pays' principle as a principle of environmental policy. According to this principle the polluters should bear the external costs of pollution. Its aim is to merge environment in economic decision making, that is, to internalise the externalities in production and consumption decisions in such a way that the prices reflect the social costs of the products.

The World Commission for Environment and Development (Brundtland Commission) *Report on Our Common Future,* 1987, stresses the need for merging economics and environment in decision-making to achieve sustainable development. According to the Commission, the need for this merger emanates from the existence of and pervasive nature of environmental externalities, institutional myopia and sectoral self-interest.

The United Nations Conference on Environment and Development (UNCED), Earth Summit, 1992 held at Rio de Janeiro, states the following objectives of environmental policy:

1. To incorporate environmental costs in the decisions of producers and consumers, to reverse the tendency to treat environment as a 'free good' and to pass these costs on to other parts of society, other countries or to future generations.
2. To move more fully towards the integration of social and environmental costs into economic activities, so that prices will appropriately reflect the relative scarcity and total value of resources and contribute towards the prevention of environmental degradation.
3. To include, wherever appropriate, the use of market principles in the framing of economic instruments and policies to pursue sustainable development (*Agenda 21,* Chapter 8, p.85).

Both the *Declaration of Rio* and *Agenda 21* endorse applications of the 'polluter pays' principle and the precautionary principle to environmental policy. While the 'polluter pays' principle makes the polluters bear the costs of pollution, the 'precautionary' principle recognises the existence of uncertainty and seeks to avoid irreversible damages *via* the imposition of a safety margin into policy.

The Tax Reforms Committee headed by Raja Chelliah (Ministry of Finance, 1992), recommended that excise taxes could be an useful instrument for internalising negative externalities in production.

In its Judgement dated 21 August, 1996 on the Public Interest Writ Petition (c) No. 914 of 1991 filed by the Vellore Citizens Welfare Forum, the Supreme Court endorsed the 'polluter pays' principle and the 'precautionary' principle. The Supreme Court, in its various judgements, refers to the 'polluter pays' principle as part of the basic environmental law of the land and requires that a polluter bears the remedial or clean up costs as well as the amounts payable to compensate the victims of pollution.

The Ninth Five Year Plan, 1999, lists one important element of environmental strategy as, "integrating environment with decision making through valuation of environmental impacts; evolving market based economic instruments as an alternative to the command and control form of environmental regulation; appropriate pricing of natural resources based on their long-term marginal cost of supply; appropriate fiscal reforms and natural resource accounting" (p.952).

Initiatives Taken by the Ministry of Environment and Forests

a. *The Policy Statement for Abatement of Pollution,* issued by the Ministry of Environment and Forests in February 1992, admits that 'the state of the environment continues to deteriorate'. It favours 'a mix of instruments in the form of legislation and regulation, fiscal incentives, voluntary agreements, educational programmes and information campaigns'. It recommends the 'polluter pays' principle, involvement of the public in decision making and new approaches for considering market choices 'to give industries and consumers clear signals about the cost of using environmental and natural resources'.

b. At the instance of the MoEF, with financial support from the World Health Organization, Mehta *et al.* (1993/1997) reviewed the theoretical and empirical literature on policy instruments for pollution control, estimated the pollution abatement cost function for Indian paper and pulp industry and considered problems in the implementation of a pollution tax.

c. The Ministry constituted a Task Force on August 1, 1995 to carry out a study to evaluate the market based instruments (MBIs) for industrial pollution abatement. The Task Force headed by the Director, National Institute of Public Finance and Policy, was asked to (1) review the regulatory and fiscal regime in industrial pollution abatement in India, (2) suggest measures to improve the effectiveness of existing fiscal instruments, (3) identify additional MBIs and examine their feasibility, and (4) develop a plan of action for introducing MBIs. Some important recommendations of the Task Force are: (1) Amendment of the current pollution control laws (or enactment of new legislation) is needed to allow explicit incorporation of MBIs. (2) In the short to medium term, greater reliance should be placed on economic penalties than in the past. In the long run, MBIs should replace criminal provision in the laws. (3) Tax concession on installation of pollution control equipments should be abolished. (4) Pollution taxes are needed for small dispersed source of emissions/effluents in accordance with the 'polluter pays' principle. In cases where emissions/effluents are difficult to measure, complementary inputs and/or outputs should be taxed. (5) Use tradable permits for firms, provided there is an adequate number of firms in the market (NIPFP, 1997).

d. The Ministry implemented, with the World Bank Assistance, India: Environmental Management Capacity Building Technical Assistance Project during 1997-2003. An important component of capacity building related to environmental economics. The objective was to increase the capacity for the application of economic principles and tools to environmental management in India across the full range of issues such as priorities setting, cost benefit analysis, alternative policies for pollution control, resource management and biodiversity conservation.

e. The Ministry, in cooperation with the World Bank and the Confederation of Indian Industry, organised a three-day International Workshop on Economic Instruments for Prevention and Control of Industrial Pollution during 13-15, June 2001. The objective of the workshop was to appraise the Indian industry and policy makers in the Government of the experiences of developing countries such as Colombia, China, Indonesia, Philippines and Thailand on the design and implementation of EIs. There was a general consensus that the time was ripe for introducing EIs for prevention and control of industrial pollution in India. The workshop recommended that (a) EIs should serve as complements to the existing standards regime and not as a substitute; (b) EIs should be used as an incentive mechanism to encourage industry to adopt environment friendly technologies; (c) preparatory work should be done on the relative costs and benefits of EIs compared with CAC instruments; and (d) involvement of the stakeholders in the design and enforcement of EIs is necessary to ensure the success of EIs in improving environmental quality; see World Bank, MoEF and CII (2002).

f. With a view to further examining the recommendations of the International Workshop, the MoEF constituted a Task Force for the design and enforcement of EIs by way of suitably framing environmental policies, laws, standards and guidelines for their introduction. At its first meeting held on November 1,

2001 Shri. Roy Paul, Special Secretary and Chairman of the Task Force, suggested that Dr. Raja Chelliah be requested to prepare a proposal on taxes on outputs and inputs causing negative externalities. This project was undertaken as the first activity of the Centre of Excellence in Environmental Economics at Madras School of Economics funded by the Ministry from April 1, 2002 in collaboration with the National Institute of Public Finance and Policy.

2

Design and Implementation of Ecotaxes

Why Ecotaxes?

Ecotaxes on pollution or pollution causing inputs and outputs can signal the costs of using environmental resources and thereby internalise the negative externalities in decision making by producers and consumers. An ecotax is a price-like instrument which assigns a price to the 'unpaid factor' of production. It can translate the 'polluter pays' principle into practice.

Even though a quantity-based instrument i.e., tradable pollution rights, can also help in improving environmental quality, there are legal and institutional obstacles to introduce such an instrument in India. Our legal system does not permit creation and assignment of property rights for environmental resources.[1] Even if the legal barrier is removed, we have to develop the institutional capacity to design and enforce a tradable permit regime. The markets may be thin and the market results may not mimic competitive market outcomes. On the other hand, our legal system permits taxes on outputs and inputs. We do have a well-developed tax administration for commodity taxes. For ecotaxes on polluting inputs and outputs, we need information on the marginal damage costs for determination of the tax rates. For direct taxes on pollution, we need reliable database on pollution, both quantity of effluent/emissions and concentrations of pollutants in emissions/effluents as well as marginal abatement costs or values of marginal damages corresponding to the prescribed standards.

Weizsacker (1992) and some other economists popularised the idea of revenue-neutral green tax reforms, that is, reforms in which environmentally motivated taxes replace income or factor taxes.

1. The Supreme Court of India interprets Article 21 dealing with the right to life as the right to clean air and water, a wholesome environment. The state is viewed as a trustee of natural resources like air and water in rivers and lakes.

There are many theoretical and empirical studies on this hypothesis. There are a few enthusiasts for such tax shifting. Mankiw (1999) argues that a 10 per cent reduction in the US income tax and a 50 cent per gallon hike in the tax on gasoline "would lead to rapid growth, less traffic congestion, safer road, and reduced risk of global warming all without jeopardising long term fiscal solvency. This may be the closest to a free lunch that economics has to offer."

An ecotax on a polluting output would increase the market price of the product and thereby decrease the level of demand and hence the production. The extent of increase in the price and the reduction in the demand depends, amongst other things, on the own-price elasticity of demand. As the level of output falls, the pollution load from the industry will also fall. An ecotax on a polluting input would raise the unit cost of production. A producer can respond to the rise in the unit cost by reducing the level of output or/and substituting an eco-friendly input for the polluting input. Either of these choices will reduce the pollution load. Imposition of an ecotax will also generate a new source of revenue to government which can be used to support activities related to environmental protection or/and reduce distortionary taxes on factors of production. These two benefits, i.e., reduction in pollution and increase in government revenue, arising from the imposition of an ecotax on polluting input and output is called double dividends.

In the Indian context, the concern is not for revenue-neutral green tax reform. The government needs resources not only for an effective implementation of the existing CAC regime, but also for undertaking many activities to arrest degradation of land, afforestation, preservation of biodiversity and prevention and control of pollution (see *Economic Survey 1998-99*). *Human Development Report 1998* estimates the annual costs of environmental degradation in India at US$ 10-13.8 billion or 4.5-6.0 per cent of GDP. A well-designed package of ecotaxes would not only improve environmental quality but also can help to some extent to generate funds to finance sustainable development.

Coverage and Design Issues

We now consider some theoretical, empirical and policy issues in introducing ecotaxes (Sankar, 2002).

Pollution may be from stationary or point sources as in the case of emissions and effluents discharged into atmosphere or water bodies by factories. The point sources may be large, medium or small in terms of size, measured in terms of output. The source-specific standards and controls regime covers only the point sources. The point sources are expected to get consent to establish and consent to operate certificates from their respective State Pollution Control Boards (SPCBs) and comply with the SPCB standards and regulations. The present system permits polluting units to discharge emissions/effluents up to certain permissible limits but *prohibits* discharges above the limits. While the cost of compliance varies with the extent of compliance, that is, the cost of reducing pollution parameters from the levels in the influents to the levels in the effluents, the penalty is unrelated to the cost of compliance. As a result, non-compliance is ubiquitous. A pollution charge varying with the extent of compliance provides a choice to the polluting units to decide on whether to comply with the standards by preventing and abating pollution and thereby avoiding the charge or make payment for the excess pollution (difference between pollution level in the effluent and the standard). Environmental effectiveness will be higher if the charges are set above the marginal abatement costs corresponding to the standards. If a tax on output or input is levied on units coming under the standards and controls regime, it will become necessary to give rebates on the ecotaxes for the complying units to motivate the partially complying units, to comply with the standards.

For non-point sources such as vehicular pollution and pollution occurring at the consumption stage, application of a direct instrument like pollution charge is not feasible; hence one has to rely only on indirect instruments such as taxes/charges on inputs and outputs. Here, the extent of environmental effectiveness would depend on the link between output/input and pollution. Only in the simple case where pollution per unit of output/input is constant, a tax on output/input will be equivalent to a pollution charge.

Pollution may be a global bad, regional bad or a local bad. In the case of global bad (pure public bad) only the aggregate pollution matters and not where the pollution is generated. Even in cases of other global bads, e.g., ozone depleting substances and biodiversity loss, developing countries have 'common but differentiated responsibilities' under the Rio Declaration. The multilateral agreements provide mechanisms for technical and financial assistance from developed to developing countries. Hence, both in terms of our national priorities and international obligations, there is no urgent need, at present, to worry about introducing ecotaxes on global bads. For this reason, even though a carbon tax is easier to design and implement and has a huge revenue potential, we do not recommend introduction of such a tax now. It is better to introduce this tax when other countries also introduce the tax so that our competitive position is not eroded.

Some pollutants are uniformly dispersed and hence ambient concentrations depend largely on the total emissions. Some other pollutants are not uniformly dispersed and hence the impact of pollution will be higher in regions where the emissions are discharged, e.g. dusts in the vicinity of cement factories, particulate matter from vehicular pollution in metropolitan cities, and effluents discharged on land affecting land quality and water quality. In such cases, we need differential rates for different regions, depending on the marginal damages. We may also need a combination of regulations and EIs (economic instruments) to reduce the pollution.

Control of air pollution, particularly pollution measured as sulphur dioxides (SO_2), nitrogen oxides (NO_x) and particulate matter (SPM), pollution in interstate rivers and marine pollution can be at the national level. A national policy can accommodate different pollution charges for different regions based on regional ambient concentrations and regional carrying/assimilative capacities. In general, local water pollution control and solid waste management can be entrusted with state level and local governments' level.

In our federal system, different layers of government have different tax jurisdictions. Our focus is on design of ecotaxes within the central government tax system mainly union excise import duty and cess. Our main thrust is on the design of taxes on outputs and inputs of polluting activities. We choose this option because there are no legal and institutional impediments to introducing such taxes. While discussing the introduction of ecotaxes, we would have to also examine the existing structure of relevant taxation to see where it would have to be modified.

An ecotax as an input is appropriate if its use in production generates pollution, as for example, use of fossil fuels in manufacturing and power generation, and of chemicals in manufacturing. An ecotax on an input based on its environmental damage can give a signal to the producers to improve the quality of the input, as for example, the use of washed coal or finding an environment-friendly substitute. The tax can be made rebatable in a VAT system if the plant has control devices to reduce pollution to the permissible level. The effect of the tax on the final price would depend on, among other things, the tax rate, the share of the input in the cost of production, the price elasticity of demand and the market structure.

An ecotax on output is desirable if pollution occurs at the consumption stage as in the case of use of petrol or diesel, plastic products, fertiliser and pesticides. The link between quantity of output and pollution may not be monotonic. For example, use of chemical fertiliser up to certain level may be beneficial but beyond the level environmental damage would occur. The same is the case with a chemical pesticide. Further, the effect of fertiliser or pesticide use on the environment depends on the manner of use and on weather conditions. As there are information problems and high transaction costs in monitoring the use at the consumption stage, a tax on output appears to be a second-best solution in several cases.

Ecotaxes can also be on observable/measurable proxies which have links with pollution. Some examples are taxes on auto fuel rather than on emissions, weight of car, fuel efficiency, engine emission factors, and age of vehicle rather than on actual emissions. Even though the links between these observable proxies and pollution are not direct and perfect, the former are preferred as tax bases because of the simplicity taxing than from the administrative point of view.

A new tax proposal must also take into consideration the existing distortions, if any, in the existing tax structure and pricing policy. As the Tenth Five Year Plan envisages eight per cent GDP growth and as the economy is being liberalised and globalised, the effects of the tax proposals on growth of industry, employment and competitiveness of the industry must also be considered. It should also be examined if there could be reduction in the rate of distortionary taxes. Another relevant factor is the availability of reliable information base for the measurement of the tax base and determination of the tax rates. Even in an industry/sector identified for ecotaxes, many practical questions do arise regarding coverage and tax exemptions. The transaction costs of designing and enforcing ecotaxes must also be kept in view. Therefore, it becomes necessary to identify the products/inputs/pollutants suitable for ecotaxes keeping in view considerations such as economic efficiency/cost effectiveness, environmental effectiveness, competitiveness, revenue potential, recycling part of the revenues to the concerned industries for adoption of clean technologies and end of pipe treatment for getting industry support. Creating awareness among political leaders, industry, labour and general public about the benefits arising from the use of EIs is also important.

International Experience

We review briefly international experience on the design and enforcement of ecotaxes on outputs and inputs and emission charges.

OECD (1997) survey reveals that in 1997 there were 284 environment related taxes and charges in the OECD countries, of which 181 were in use in the European Union and 103 in non-European OECD countries. Among the European Union countries, Denmark, Finland and Sweden had more than 20 environmental taxes/charges. Among the non-European countries Norway had a maximum of 22 taxes/charges followed by Australia (15) and the USA (14). The break-up of the taxes itemwise is given in Table 2.1. Further details are given in Annexure A-2.1.

Table 2.1

Environmental Taxes in OECD Countries

Taxes Charges on	*OECD Countries*	*EU Countries*
Motor Fuels	61	40
Other Energy Products	35	26
Vehicles	35	20
Agricultural Input	6	4
Other Goods	25	15
Direct Tax Provision	28	17
Air Transport	14	7
Water	39	24
Waste Disposal and Management	41	28
Total	**284**	**181**

Source: OECD 1994, 1996 and 1997.

It may be noted that taxes/charges on motor fuels and other energy products accounted for more than one-third of such taxes/charges in the OECD countries; vehicles taxes accounted for 12 per cent of the taxes/charges. Table 2.2 gives information about product charges in European countries. Information about the levies imposed on fertiliser in four European countries and the USA are given in Table 2.3. Table 2.4 gives particulars regarding levies imposed on pesticides in Norway and Sweden. A summary of features regarding vehicle tax in several countries is given in Chapter 4.

Table 2.2

Product Charges in Select Countries

Country	Product	Rate	Impact
Finland	Beverage container	ECU 0.07 for glass or metal & 0.20 for other material.	Revenue of 1.65 in 1985. Strong incentive impact.
	Lubricant oil	ECU 29/tonne.	Revenue: ECU 3 mm
	Carbon tax	ECU 5 per tonne of carbon: all fossil fuels except motor fuel. Tax on motor fuel was increased in 1990.	Revenue was approximately ECU 61 mm in 1991. In additional ECU 173 mm was raised from higher motor fuel taxes. Tax indexed to inflation.
Denmark	Beverage container	ECU 0.06 to 0.28 per container.	ECU 35 mm in 1987 strong incentive impact.
Norway	Beverage container	ECU 0.08-0.52 per container.	ECU 11 mm in 1987: too low for incentive impact.
	Mineral oil	General tax of ECU 2.7/100 litres. Additional tax of 1.9/1000 litre, if sulphur content is 0-25 per cent. ECU 3.9, if content is 0.25-0.5 per cent. ECU 5.85 per cent if content is 0.5-75 per cent.	ECU 27 mm in 1988: tax too low to have incentive effect.

Contd. ...

Contd. ...

Country	*Product*	*Rate*	*Impact*
	Fertiliser & pesticide	Rate: ECU 0.13/kg of nitrogen and 0.25 per kg of phosphorous in 1992.	In 1992, revenue was ECU 19 mm from fertiliser tax and 2.5 mm from pesticides tax. Revenue used to finance environmental programmes in agricultural sector.
	Batteries	ECU 0.26 in 1989, ECU 3.27 per kg of batteries with Hg or cadmium.	
	Carbon tax	ECU 33/tonne of carbon introduced in 1991 (as in Sweden).	ECU 5 mm from general batteries & ECU 2.4 mm from cadmium batteries.
Sweden	Beverage container	ECU 0.014 to 0.035 by size.	
	Mineral oil	ECU 17.76/m^3 of which ECU 1.26 is for measures against acidification.	
	Fertiliser & pesticides	ECU 0.08/kg of nitrogen & 0.04/kg of phosphorous. Pesticides ECU 0.64/kg active ingredient & ECU 4/ha.	ECU 24 mm in 1992 from fertiliser and pesticides tax. Revenue use for for agricultural & forestry programmes.
	Carbon tax	ECU 33/tonne of carbon–introduced in 1991.	Expected revenue in 1992 ECU 1.46.
France	Lubricant oil	ECU 6/tonne	Revenue: ECU 4 mm-revenue earmarked and used to enforce proper collection and encourage recycling less use of oils.

Contd. ...

Contd. ...

Country	*Product*	*Rate*	*Impact*
Germany	Lubricant oil	ECU 96/tonne	Revenue: ECU 63 mm–oil disposal reduced.
The Netherlands	Lubricant oil replaced by general fuel charge	ECU 6/1000 litres	ECU 1 mm revenue used to encourage recycling & reduce use of oils.
Italy	Lubricant oil	ECU 3/tonne	ECU 2 mm in Germany, Netherlands and Italy revenue is used to enforce proper collection and encourage less use of oil and recycling.
	Plastic bags	Tax of ECU 0.06/bag costing 0.012 to make or bags with bio degradability of < 60 per cent	Not known

Source: Opschoor and Voss (1989), OECD, 1992.

Table 2.3

Levies Imposed upon Fertilisers in Select Countries

Country	*Charge Base*	*Rate*	*Per Cent of Price*	*Revenue Spending*
Austria	Nitrogen content	ECU 0.31/kg		Subsidies, environmental expenditure.
	Phosphorous content	ECU 0.18/kg		
	Potassium content	ECU 0.09/kg		
Finland	Nitrogen content	ECU 0.41/kg	5-20	Agricultural subsidies, general budget.
	Phosphorous content	ECU 0.27/kg		
Norway	Nitrogen content	ECU 0.13/kg	19	General budget.
	Phosphorous content	ECU 0.24/kg	11	
Sweden	Nitrogen content	ECU 0.07/kg	10	Subsidies, environmental expenditure.
	Phosphorous content	ECU 0.14/kg		
USA		ECU 0.07–1.11 per tonne	≤2.5	Environmental expenditure.

Source: Adapted from: *The Fiscal System and the Polluter Pay Principle*, Barret *et al.*, (1997), p.61.

Table 2.4

Levies Imposed on Pesticides in Norway and Sweden

Country	*Levies Imposed on Pesticides*	*Revenue Spending*
Norway	13 per cent on whole sale price	General budget
Sweden	ECU 0.9 per kg active ingredient ECU 3.6 per treated hectare	Environmental expenditure

Source: OECD, 1994, p.75.

We summarise the novel features of a few specific taxes focussing on problems such as multiple objectives, coverage, incentives and penalties, and earmarking of revenues. Bregha and Moffet (1995) study the fuel conservation tax introduced in Ontario, Canada in 1991. This tax scheme has multiple objectives: (a) environmental protection, (b) energy conservation, and (c) revenue generation. Fuel conservation of a vehicle is based on highway fuel consumption ratings, that is number of litres per 100 km, classified under many classes with rebate for the lowest class of less than 6 and tax rate increasing sharply from C$ 75 for the second class of 6.0 to 7.9 to C$ 700 for the last class of 15 per cent or more. The rebate for the lowest class is a gas sippers subsidy while the highest rate for the largest class is a gas guzzlers tax. Even though there was opposition to the tax scheme due to the recession, the authors argue that this scheme created a political constituency for green taxes and the need to reconcile potentially competing environmental and fiscal objectives.

Schnutenhaus (1995) outlines features of the German Tax Programme. They are: (a) tax differentiation in favour of unleaded petrol, (b) tax differentials in favour of low emission vehicles, with higher rates for diesel engine vehicles than for petrol engine vehicles, (c) tax exemptions for cars with catalytic converters, and (d) incentives (cash payment) for retrofitting existing cars with catalytic converters.

Olivecrona (1995) presents some interesting features in the design and implementation of nitrogen oxide charge on energy production in Sweden. The objective of the tax was to contain the NO_x emissions in 1994 at 1987 level. Even though NO_x emissions result in combustion plants for energy production in paper and pulp, chemical and metal industries, and waste incineration plants, as these emissions are diffuse and the costs of measurement are too high, they were not subject to the tax. Even in the energy sector, only combustion plants with capacity of 10 MW or more and an annual energy production exceeding 50 giga watt-hours were liable for the charge. While the cost of reducing one kg of NO_x emitted varied between 3 and 80 SKr, the charge was fixed at 40 SKr. To facilitate the acceptance of the charge, the income from the charge was distributed to liable plants in relation to their energy production. This refund stimulated the plants even more efficient. For smooth implementation, each plant must declare the amount of NO_x charge payable along with a calibration certificate of the monitoring equipment from an independent examiner. The Swedish Environmental Protection Agency examines and in some cases carry out audit. Not only the net benefit of this programme was positive but this experience strengthened Swedish position in the international market for NO_x reducing techniques.

China has been implementing a pollution levy system (PLS) for the last 15 years. It adopted the 'polluter pays' principle. In the beginning the rates were kept low and the levies were only on certain pollutants. In the new PLS introduced in 2000, the tax base is wider and comprehensive covering all pollutants from all sources. This system has lower rates for emissions below the official standards and higher rates for emissions above the standards. It allows for annual escalation of the rates to account for inflation and growth. It has adopted the pollution equivalent concept to provide an administratively feasible basis for the system. By this concept, each pollutant is rated according to its harmfulness and assigned a pollution equivalent index value greater than or less than the reference emission which is assigned a value of 1. Around 30 per cent of the revenue raised was earmarked for supporting environmental administration at all levels. From the balance, three funds were created—municipal fund to receive 30 per cent, provisional fund to receive 15 per cent and national fund to receive 55 per cent. Bohm (1996) considers the PLS reform a bold and comprehensive effort to stop China's environmental deterioration and to place the pollution control enterprises on a solid footing for the future.

For the experience of other developing countries in the introduction of pollution charges/taxes, see O'Conner (1998) and World Bank (2000).

We summarise a few lessons from the experiences of other countries in designing and implementing the pollution charges/taxes. First, it is necessary to develop an enabling legal and institutional capacity to design and enforce the taxes/charges. Second, the trade-offs between environmental benefits and other objectives must be assessed and political backing must be ensured. Third, it is desirable to involve the stakeholders—polluters, pollutees, NGOs and others in the design of the taxes/charges. Fourth, the charge/tax base and the units to be covered must be arrived at keeping in view the transaction costs of implementation. Fifth, lack of information about marginal abatement costs/marginal environmental damages need not deter the introduction of such charges or taxes. In fact, many countries set the actual rates far below the marginal abatement costs/damage values to popularise the idea of pollution charges/taxes. As in the Chinese and Columbian cases the rates can be revised over a period of time taking into consideration the relation between the rates and environmental effectiveness. Sixth, the introduction may be done on a pilot basis and sufficient time may be given to the affected parties to prepare for the change. Seventh, earmarking of revenues from environmental taxes/charges for development of clean technologies, erection of effluent treatment plants and pollution control devices, environmental awareness campaign and capacity building in environmental administration have public support and industry acceptance.

Scope of the Study

This study is confined to levy of taxes/charges/fees on inputs and products of a few polluting industries and exploration of the feasibility of introducing emission fees/charges on a pilot basis. As mentioned earlier, the justification for introducing taxes/charges/fees on polluting outputs and inputs is that we do possess the legal and institutional capacity to design and enforce the taxes/charges/fees. The taxes/charges/fees signal the social costs of pollution to the polluters and also generate revenues for government. However, even the design of taxes/charges/fees require preparatory work on decisions such as the tax base and its coverage, determination of the rates, uses of the revenues, creation of awareness among all the stakeholders about the need for such taxes/charges/fees, and assessments of the trade-offs between environmental and other objectives. The proposed tax scheme is complementary to the existing pollution control regime.

Each chapter gives some background of the industry, its environmental problems, coverage under the tax net, determination of the tax base and tax rate, the manner in which the tax is to be included in our tax system, the use of revenues, and the likely effects. In the determination of the rates we have kept in mind both the existing tax and price distortions and the environmental harm.

Annexure A-2.1

Environmental Taxes in OECD Countries

The OECD member countries are increasingly employing environmental taxes to improve their environmental performance. Table A-2.1 gives some of the environmental taxes in use in the OECD countries:

Table A-2.1

Environmental Taxes on Industries in OECD Countries

Country	*Tax*
Denmark	Differing tax rates on petrol stations depending on the vapour recovery system used.
Norway, Denmark	Tax on tetrachloroethylene and chlorinated solvents.
Sweden, Switzerland	Tax on volatile organic compounds.
Norway, Sweden, Finland and Denmark	Tax on carbon dioxide emissions.
Austria	Tax on electricity usage.
Italy	Carbon monoxide on mineral fuels.
The Netherlands	General fuel tax.
Belgium	Tax on some energy products.
Belgium, Canada, Denmark, Finland and Norway	Pesticides.
Germany	Taxes to promote energy efficiency, industrial reform and to fund renewable energy.
The Netherlands	Surplus phosphate and surplus nitrogen.
U.S, Norway, Sweden, Denmark	Fertiliser.
Denmark, Finland, Hungary, Italy and Norway	Lubricant oil.

Source: http://www.audit.vic.gov.au/index.html and OECD publications.

Most of the OECD countries have enunciated a comprehensive environmental tax reform. Environmental taxes seem to be more popular in Europe particularly in Northern Europe as shown in Table A-2.2.

Table A-2.2

Environmental Taxes in European Countries

Tax	*Country*
Nitrogen Oxide Tax	Sweden, Spain, France
Water Abstraction Tax	Netherlands, Denmark
Wastewater	Denmark, Netherlands and Germany
Pesticides	Sweden, Denmark and Belgium
Fertiliser	Netherlands, Finland, Sweden and Austria
Landfill	France, Britain and Austria
Packaging	Finland, Denmark and Sweden
Batteries	Italy, Belgium and Hungary

Denmark, Finland, Germany, Italy, the Netherlands, Norway, Sweden and the United Kingdom implemented explicit ETR through a package of measures. In all these countries the Environmental Tax Reform (ETR) packages have tended to reduce the burden placed on labour by reducing social security contributions paid by employers. The focus of the new higher green tax in the ETR has been the energy sector, due to the compelling need to arrest global warming as well as the revenue potential of the energy tax.

Revenue Potential of Environmental Taxes

A study of 19 developed countries by OECD shows that the share of revenues from green taxes in gross domestic product was less than 2 per cent in 1995 for each of the 19 countries studied. On average, environmentally-related taxes in OECD countries came to about 2.7 per cent of GDP in 1998 and 7.2 per cent of the total tax revenue. Denmark's 'green' taxes accounted for 5 per cent of GDP in 1998, up from little more than 4 per cent in 1994, but while Turkey more than doubled its green tax level to more than 3 per cent of GDP over the four years, the United States, Germany and France actually reduced their green tax ratio. Greece, the Netherlands, Norway, and Portugal were slightly below this level while Mexico and the United States had the lowest ratios for the group, at just about 1 per cent.

Revenue Potential of Taxes on Fuel

A study by OECD reports that the maximum revenue from green taxes is from the tax on unleaded petrol for the 19 industrial countries, accounting for nearly 40 per cent of the total revenue raised in 1995, followed by motor vehicles, accounting for more than 20 per cent. Taken together, taxes on petroleum, diesel fuel and the sale or use of motor vehicles raised more than 91 per cent of all environmentally-related revenue in the countries covered by OECD study. The OECD, Environmental Outlook, observes that applying a tax on fuel use and removing all energy subsidies could reduce carbon dioxide emissions by 25 per cent by 2020 in OECD countries and 11 per cent worldwide compared with a business-as-usual scenario. Most of the countries have introduced a tax on fuel and energy as a step towards reducing CO_2 emissions. Norway, Sweden, Denmark, Finland, Italy and the Netherlands have introduced CO_2 taxes. In most of the OECD countries, leaded petrol has disappeared from use partly due to higher tax and partly due to regulation. Likewise differential tax rates are applied on alternative fuels and fuel oils in many countries.

- In Denmark, the tax rate on petrol differs depending on whether the petrol station that sells it is equipped with a vapour return system.
- Norway levies special sulphur tax on diesel with a sulphur content of above 0.005 per cent. This is in addition to the ordinary SO_2 tax rate.
- In Sweden, different rates of taxes are levied on the three cases of petrol and diesel depending on the sulphur and benzene content, density and distillation interval.
- UK applies a lower tax on low sulphur diesel and petrol.

Economic Impact of Environmental Taxes

Andrew J. Hoerner and Benoit Bosquet of Center for a Sustainable Economy, assessing the economic impact of ETRs in the European countries, based on 44 studies containing 104 distinct simulations of environmental tax reform, observe that when the revenues of environmental taxes are used to reduce other distorting taxes, the economic outcome is better than if those revenues are not so distributed, in terms of impacts on both employment and GDP.

Seventy-eight per cent of the simulations predict that ETR will create employment. The best results in terms of employment are obtained when recycling occurs through cuts in social security contributions, with 86 per cent showing positive employment results, as opposed to, e.g., 35 per cent positive results for income tax cuts. (Hoerner and Benoit, 2001). Their study suggests that ETR packages that recycled a portion of the environmental tax revenues to finance energy efficiency or renewable energy improvements are more likely to result in positive employment and GDP impacts. Table A-2.4 gives a summary of the ETR packages in some of the European countries that are OECD members.

Table A-2.3

Environmental Taxes on Transport in OECD Countries

Country	*Tax*
Austria, Finland, Denmark, Norway, Sweden, United Kingdom	Diesel taxes.
Austria, Germany, Norway, Denmark	Differentiated vehicle taxes—adjusted according to the emissions of the vehicle.
United States of America	Tax on sale of new vehicles with a fuel efficiency less than 22.5 miles per gallons—introduced in 1978-tax ranges from $US1000 to $US7700 per automobile. Minivans, trucks and sport utility vehicles are exempt.
Japan	Differentiated annual vehicle registration fees—older vehicles (with higher emissions) pay more.
Austria, Ontario (Canada)	Environmental taxes on motor vehicles.
Austria	Tax on registration, depending on fuel efficiency. The tax is higher for diesel vehicles.
Japan, Ireland, United States of America, Canada	Tax on aviation fuel for domestic flights.
Norway	Tax on aircraft passenger seats, includes international flights.

Source: *The United States Experience with Economic Incentives for Promoting the Environment*, U.S. Environmental Protection Agency, EPA-240-R-01-001, January 2001 and *Environmentally Related Taxes OECD Countries-Issues and Strategies*, OECD, 2001.

Table A-2.4

ETR Packages in Select OECD Countries

Country	*Taxes Cut or Items Funded*	*Taxes Raised on*	*Magnitude*
Sweden (1990)	• PIT • Energy taxes on agriculture • Continuous education	• CO_2 • SO_2 • Various	2.4 per cent of total tax revenue.
Denmark (1994)	• PIT • SSC	• Various (gasoline, electicity, water, waste, cars) • CO_2 • SO_2 • Capital gains	Around 3 per cent of GDP by 2002, or over 6 per cent of total tax revenue.
Netherlands (1996)	• CPT • PIT • SSC	• CO_2	0.3 per cent of GDP in 1996, or around 0.5 per cent of total tax revenue.
United Kingdom (1996)	• SSC	• Landfill	Around 0.1 per cent of total tax revenues in 1999.
Finland (1997)	• PIT • SSC	• CO_2 • Landfill • Corporate profits	0.3 per cent of GDP as of March 1999, or around 0.5 per cent of total tax revenue.
Norway (1999)	• PIT	• CO_2 • SO_2 • Diesel oil	0.2 per cent of total tax revenue in 1999.
Germany (1999)	• SSC • Renewable energy	• Petroleum products	Around 1 per cent of total tax revenue in 1999.
Italy (1999)	• SSC	• Petroleum products	Less than 0.1 per cent of total tax revenue in 1999.

Source: Hoerner and Benoît (2001). *http://www.rprogress.org.*

Observations on the Environmental Taxes

Positive and significant environmental impacts of environmental taxes have been observed by some studies: A study by ECOTEC in association with CESAM, CLM, University of Gothenburg, UCD and IEEP (CR) on environmental taxes in EU, reports the following positive environmental impacts of green taxes specific taxes in select EU (also OECD) countries:

- Between 1992 and 1998, the use of refunded NO_x charges in Sweden doubled the number of plants liable to pay the tax besides reducing the NO_x emissions by four per cent.
- Reductions in CO_2 emissions due to emission tax—emissions from stationary combustion plant in Norway decreased by 21 per cent in just four years.
- In Sweden, SO_2 emissions were expected to decrease by 80 per cent of 1980 levels, by 1999, since the introduction of sulphur tax in 1991.
- The Danish water abstraction tax complemented by environmental awareness programmes, achieved a 13 per cent reduction in water consumption since the tax was introduced in 1994.
- The Dutch and German wastewater levies have helped to reduce COD wastes in water significantly.
- The fertiliser levies in Austria, Sweden, the Netherlands and Finland have recorded reductions in fertiliser use. The pesticide tax too has been effective in reducing the use of harmful pesticides in Sweden and Denmark.
- The landfill taxes in UK and Austria have increased recycling and led to minimisation of industrial and commercial wastes.
- The French NO_x tax and the Austrian fertiliser and manure tax has had limited positive impact.

Some studies observe that the role of environmental taxes in the OECD countries is restricted (Markandya, 1993). Information on emission charges, user charges, product charges, and tax differentiation for the OECD countries is summarised in Tables A-2.5, A-2.6, A-2.7 and A-2.8. The following points are observed from the table by Markandya.

1. In almost all OECD countries the environmental taxes are accompanied by some kind of restriction either on permissible level of emission or on technology to be installed.
2. The tax base is not clearly defined and in most cases is not in terms of damages inflicted by the pollution.
3. Revenue incentive appears to be the predominant motive even though incentive goals are significant in exceptional cases like tax on beverage container, plastic bags, etc.
4. Though taxes claim to be based on 'polluter pays' principle, in practice the claim appears objectionable since the revenue collected is used to install pollution control equipments in firms from which the tax revenue has been raised.
5. The share of environmental taxes in total taxes remain an insignificant proportion and hence there is considerable scope for restructuring the tax system to reduce some of the sources of other taxes and increase the environmental taxes.
6. The level of authority for administering the taxes vary. While product charges and charges on air pollution are collected at national level, water effluent charges, solid waste charges and user charges are collected at the local or regional level (Markandya, 1993).

The charges are set at too low levels in almost all the OECD countries except the Netherlands' water effluent tax that exceed the marginal abatement tax achieving significant reductions in effluent generation. This high rate has been possible because of the political necessity of hydraulic control in the Netherlands that has made strict systems acceptable to Dutch industry.

Conclusion

It may be concluded that there is ample scope for further taxation of environmental services in OECD countries. It is observed that environmental taxes perform better when they are part of a more integrated approach as in the case of the Danish pesticide tax. Designing of the tax is very crucial—the Dutch levy on nutrient surpluses, the Swedish refunded NO_x charge and the Danish packing tax based on life cycle impacts and Norway's pesticides tax are some good examples of well designed tax. In most cases the environmental effects of the tax are compromised by the exemptions that accompany the tax. It is feared that the green taxes may harm international competitiveness. Some ecotax leaders have therefore granted tax exemptions to some heavy emitters of greenhouse gases—although such exemptions will impair the double dividend effect of the tax. For example, Finland has exempted the electricity sector from the carbon tax. Denmark 'recycles' the yield from emissions taxes to industry and, in some cases, has replaced taxation with voluntary agreements. And, Norway has been considering scaling back its carbon tax in response to falling investment in the oil sector. So, policy trade-off between environmental goals and industries' competitiveness considerations have played a major role in the way green taxes are actually designed and used (Norregaard and Reppelin-Hill, 2000).

Table A-2.5

Environmental Taxes in OECD Countries

Country	*Structure/ Description*	*Use of Revenue/ Revenue Impact*	*Incentive Impact*	*Other Regulations and Activities*	*Other Issues*
Air					
France	ECU 19/Tonne for SO_2 plants with discharges of 2500 tonnes or generation capacity of over 50 MW.	Yearly revenue = ECU 13 mm. Ninty per cent used for subsidies for control equipment; 10 per cent for technological innovation.	None observed. Required level of tax for incentive effects would be much higher.	Several technology directives apply.	Only 400 plants affected. Distributional impact small. National level tax.
Sweden	ECU 4000/Tonne of sulphur on coal and peat from 1992.	Anticipated revenue is ECU 66 mm.	Not known but expected.	Yes.	If control measures applied all or part is rebated. National level tax.
Water					
France	Charges cover suspended matter; BOD/COD, soluble salts, ammonia/nitrogen, phosphorous.	Yearly revenue in 1989=ECU 240 mm. Used to finance operating expences of river basin agencies and subsidise pollution abatement equipment.	None intended but reductions in organic waste observed. Cost saving is about 12 per cent over direct regulations.	Restrictions on maximum releases exist and are important.	Impact has been eroded by inflation. Financing of pollution abatement tied to revenues from this charge. Regional tax with funds going to river basins authorities.

Contd. ...

Contd. ...

Country	*Structure/ Description*	*Use of Revenue/ Revenue Impact*	*Incentive Impact*	*Other Regulations and Activities*	*Other Issues*
Germany	In 1990 charge was ECU 24.4 per unit of toxicity. Toxicity defined by pollutant e.g. for COD one unit equals 50 kg oxygen. A threshold is defined for each pollutant. No charge below threshold.	ECU 135 mm in 1986. Revenues used for administration and subsidising pollution abatement. Fifty per cent used up in administration.	Generally believed to be small. Improvement in water quality attributed to strict controls and low limits. Perhaps there was more incentive effect in earlier stages when reductions to meet minimum standards resulted in rebates.	Maximum releases restricted. Significant rebates on pollution abatement equipment. If state of the art then 75 per cent discount on charges. If reduction in effluent is more than 20 per cent then full cost is deductible from charge.	Charges announced in advance to 2000. Exemptions on grounds of hardship offered. Estimated reductions in costs relative to uniform charges was 33 per cent (based on data 1985). A regional level tax.
The Netherlands	BOD/COD charge for releases into 'State Waters'. Pollution charges for release into wastewater collection systems with charges for small firms based on average releases for the technology used, and on actual releases for large firms. Rates in 1985 were ECU 12.3/34 per pollution unit.	Revenues in 1985 ECU 473 mm. Management costs were only 2 per cent for state charges and 4-9 per cent for other charges. Abatement expenditures were financial from charges.	Believed to be strong. Between 1969 and 1986 estimated reduction in effluent was 64 per cent. Gain was greatest for organic matter. For heavy metals gains came from regulations. Also negotiations were important.	Discharge permits issued by water boards were important.	Heavy metal charges difficult to enforce. No charge on phosphates (political reasons). Some evidence that too much abatement resulted as charge revenues had to be allocated to abatement. Tax both at water board level and state level.

Contd. ...

Contd. ...

Country	*Structure/ Description*	*Use of Revenue/ Revenue Impact*	*Incentive Impact*	*Other Regulations and Activities*	*Other Issues*
Waste					
Belgium	ECU 0.02/2.15/Tonne or M3 depending on type of waste and treatment before dumping. Incinerated waste has a lower rate. Recycled materials exempt.	Revenues used for general purposes.	Not known.	Restrictions on waste dumped exist.	
Denmark	ECU 5.2/Tonne 'Harmless' waste, such as painted wood, straw, demolition waste exempt.	Revenues used	Not known.	Restrictions on waste	National level tax.
France	Chemical waste, ECU 0.8 m in 1985.		Low.		High administrative costs. Now similar tax in Belgium.
The Netherlands	Surplus manure charge based on phosphate content of manure beyond what can be dumped on farmland. Above 125 kg has a charge ECU 0.25/kg to 20 kg and ECU 0.50/kg over that.	Goal is revenue raising. Estimated revenue ECU 19 mm in 1985. Funds are used for research and development.	Believed to be strong. Between 1969 1986 estimated reduction is 64 per cent. Gain was greatest for organic matter. For heavy metals gains came from regulations. Also	Discharge permits issued by water boards were important.	Heavy metal charges difficult to enforce. No charge on phosphates (political reasons). Some evidence that too much abatement resulted as charge

Contd. ...

Contd. ...

Country	*Structure/ Description*	*Use of Revenue/ Revenue Impact*	*Incentive Impact*	*Other Regulations*	*Other Issues and Activities*
			negotiations were important.		revenues had to be allocated to abatement. Tax both at water board level and state level.
Noise					
France	Fee ranged from 0.9 to times avarage landing fee depending on noise levels.	ECU 6 mm in 1985, of airport income. Some use for insulation but limited on legal grounds.	Not effective.	Yes, several on landing times, etc.	Only at Paris airports. Cancelled in 1987 on legal grounds.
The Netherlands	For aircraft, based on noise characteristics. For industry, based on noise of operations.	ECU 4.4 mm in 1985 used for insulation. ECU 10 mm.	Unlikely. Not known.	Yes, several on landing times, etc. Yes, several.	Strong objections by airlines. System complex and difficult to enforce. Replaced by general fuel charge.
Switzerland	Charge is doubled every 10dB noise increase.	Approximately 10 per cent of landing fees, about ECU 4 mm.	None observed.	Yes.	Now extended to small aircraft.

Table A-2.6

Environmental Charges

Country	*Structure/ Description*	*Use of Revenue/ Revenue Impact*	*Incentive Impact*	*Other Regulations and Activities*	*Other Issues*
User Charges					
Water collection several countries	All countries charge for water collection but only in some is pollution load reflected in the charge-Belgium (firms); Italy (household & firms); UK (firms); USA (firms).	Charges vary per kg of COD ECU 0.49 (Germany) to 0.91 (UK). They cover operating and capital costs but not always on a marginal cost basis.	Depends on the extent to which charges reflect pollution load and what other alternatives exist to get and of waste (see text).	Restrictions on amounts and types of waste are applied widely.	Relation to effluent charges is weak. Charges are generally based on average costs of treatment, although some countries have greater differences according to waste (see structure column).
Product Charges					
Beverage Containers					
Finland	ECU 0.07 for beer 0.58 for glass or metal 0.20 for other materials.	ECU 1.65 mm in 1985.	Strong shift to returnable containers for glass but not plastic bottles which remain a problem. Strong incentive impact. Too low for incentive impact.		Revenues go to public budget in all these cases.
Denmark	ECU 0.06 to 0.28 per container.	ECU 35 mm in 1987.	Strong incentive impact.		

Contd. ...

Contd. ...

Country	*Structure/ Description*	*Use of Revenue/ Revenue Impact*	*Incentive Impact*	*Other Regulations and Activities*	*Other Issues*
Norway	ECU 0.08 to 0.52 per container.	ECU 11 mm in 1987.	Too low for incentive impact.		
Sweden	ECU 0.014 to 0.035 by size.				
Lubricant Oils					
France	ECU 6/tonne	ECU 4 mm.	20 per cent motor oil collected.	Several regulations on collection and disposal of oil.	Charge is earmarked and used to enforce proper collection and encourage recycling and less use of oils. In Netherlands replaced by general fuel charge.
Finland	ECU 29/tonne	ECU 3 mm.	Better collection resulted.		
Germany	ECU 96/tonne	ECU 63 mm.	Oil disposal reduced.		
Italy	ECU 3/tonne	ECU 2 mm.	Oil collected doubled.		
The Netherlands	ECU 6/1000 Litres	ECU 1 mm.	Impact not known but likely to be small.		
Mineral Oils					
Norway	General tax of ECU 2.7/100 litres. Additional tax of ECU 1.9/1000lt. if sulphur contents is 0-0.25 per cent; ECU 3.9 if content is 0.25-0.5	ECU 27 mm estimated for 1988.	Tax is too low to have an incentive impact.		This tax is not really an environmental tax in the sense that the polluter is paying for the damage. Revenue is collected from one source and used for

Contd. ...

Country	Structure/ Description	Use of Revenue/ Revenue Impact	Incentive Impact	Other Regulations and Activities	Other Issues
	per cent; and ECU 5.85 if content is 0.5-75 per cent				a range of environmental expeditures.
Sweden	ECU 17.76/M3 of which ECU 1.26 is for measures against acidification.	Eighty-five per cent of costs come from acidification tax. Rest used for general environmental measures.	Not primary concern.		
Feed Stock					
USA	ECU 0.1-0.7 per barrel of petroleum. ECU 0.02 -0.09 per tonne of base chemicals and derivatives.	ECU 487 mm from first tax; ECU 296 mm from second. Revenues used to clean up hazardous waste sites.	Not concerned with incentives as funds clean past pollution.		Not really an environmental tax as polluter is not paying for the damage.
Fertiliser and Pesticides					
Norway	Fertiliser tax was ECU 0.13/kg of N and 0.25/kg of P in 1992.	Revenue in 1992 was ECU 19 mm from fertiliser tax and 2.5 mm from pesticide tax. Aim is revenue raising.	Not expected.		Pesticide taxes amounted to 13 per cent of purchase Price. All revenues were used to finance environmental programmes in agricultural sector.
Sweden	ECU 0.08/kg of nitrogen	ECU 24 mm in	Not expected.		Revenues used for

Contd. ...

Contd. ...

Country	*Structure/ Description*	*Use of Revenue/ Revenue Impact*	*Incentive Impact*	*Other Regulations and Activities*	*Other Issues*
	and 0.04/kg of phosphorous. Pesticides ECU 0.64 /kg active ingredient and ECU 4/ha.	1992 from fertiliser and pesticide taxes.			environmental, ag. and forestry programmes.
Austria	Low level of tax introduced in 1986.	Raised about ECU 66 mm.	Although low use seems to have been Affected.		Revenues earmarked.
Batteries					
Norway	ECU 0.26 in 1989, and ECU 3.27/kg of batteries with mercury or cadmium.	ECU 5 mm from general batteries and ECU 2.4 mm from cadmium batteries.	Not known.		Should reduce demand especially for the more highly taxed batteries. National tax.
Plastic Bags					
Italy	Tax of ECU 0.06/bag costing 0.012 to make, on bags with biodegradability of less than 60 per cent.	Not known.	Not known.	Yes, at disposal.	Charge encourages less use and more recycling.

Table A-2.7

Carbon Taxes

Country	*Structure/ Description*	*Use of Revenue/ Revenue Impact*	*Incentive Impact*	*Other Regulations and Activities*	*Other Issues*
Finland	In 1991 tax was levied at ECU 5 per tonne of carbon on all fossil fuels except motor fuel. Tax on motor fuel was increased separately in 1990.	Revenue from this tax was approx. ECU 61 mm in 1991. In addition ECU 173 mm was raised from higher motor fuel taxes.	Primary goal is revenue tasting.		Tax is indexed to inflation. Levied at point of import or production. Carbon tax amounts to 5 per cent of all taxes on energy products. Higher motor fuel taxes amount to about 15 per cent of such taxes.
Sweden	Tax of about ECU 33/ tonne of carbon introduced in 1991 on all fossil fuels in 1991.	Expected revenue in 1992 is ECU 1.4 bm.	Not known but could be substantial.	tax reform.	Part of major
Norway	Similar carbon tax in Norway.	Not known.	Not known.		
Other Air Taxes					
Sweden	ECU 5300/tonne on NO_2 from large furnaces with production of over 50 GWh.	Not known.	Not known.		

Table A-2.8

Tax Differentiation

Country	*Structure/ Description*	*Use of Revenue/ Revenue Impact*	*Incentive Impact*	*Other Regulations and Activities*	*Other Issues*
Tax Differentation-Leaded/ Unleaded Fuels					
Denmark	ECU 0.045 per litre	There is a loss of revenue as a result of the lower taxes on unleaded fuels. Estimates not available.	Some evidence for positive effect. Market share of unleaded gasoline is going up. In Germany from 11 per cent in 1986 to 28 per cent in 1987. Similar changes in other countries. However, this may not be due to the tax. Gasoline dealers are narrowing the tax differential in some cases.		
Finland	ECU 0.047 per litre				
Germany	ECU 0.030 per litre				
The Netherlands	ECU 0.071 per litre				
Norway	ECU 0.041 per litre				
Sweden	ECU 0.022 per litre				
Switzerland	ECU 0.029 per litre				
UK	ECU 0.070 per litre				
Car Prices					
Several countries	The cleaner the car the greater the tax advantage. Finland, Germany, The Netherland, Norway, Sweden and the UK have some car tax differential.	As above, this is a case where there is a loss of revenue to support an envir-mental damage.	Little information is available to date. Percentage of cars meeting higher standards is increasing.		Lower taxes are based on size of car, and year of purchase. Meeting is intended as a temporary one, until all cars meet the pollution standards.

Contd. ...

Contd. ...

Country	*Structure/ Description*	*Use of Revenue/ Revenue Impact*	*Incentive Impact*	*Other Regulations and Activities*	*Other Issues*
Various Tax Allowances					
Several countries	There are various tax allowances permitted in OECD countries for investments in pollution abatement equipment. There are not covered here as they effectively amount to subsidies.	All such schemes imply a cost in term of revenue to the government.	Incentive effects vary, but the intention is clearly to encourage a reduction in pollution.		

3

Coal

Introduction

As per Geological Survey of India, the estimated coal reserves of the country down to a depth of 1,200 metre, stood at 208.752 billion tonnes as on January 1, 1999, (CPCB, 2000: 1). India's share in world coal reserves is 2.7 per cent. About 83 per cent of the reserves are non-coking coals. The non-coking coal reserves are graded into three categories. About 16 per cent of non-coking coal is of superior grade with 5800 kcal/kg, 13 per cent intermediate grade with 5000 kcal/kg and the remaining 71 per cent inferior grade with 4000 kcal/kg (CPCB, 2000; Table 3.1).

Production

Of the 588 mines in India in 1999, 355 were underground and 233 were open cast mines, but open cast mines accounted for 75 per cent of production. The production of coal in India increased from 70 million tonnes at the time of nationalisation in the early 1970s to 310 million tonnes in 2000-01. Now, India ranks third amongst the coal producing countries in the world. Currently, 98 per cent of the coal is produced by the nationalised coal companies, of which 88 per cent is produced by Coal India Limited (CIL) and 10 per cent by the Singareni Collieries Company Limited (SCCL). As such, there is no competition in the industry, although Tata Iron and Steel Company (TISCO), Indian Iron and Steel Company (IISCO), Damodar Valley Corporation (DVC) and others produce small quantities of coal (See Table 3.1).

Coal Demand Forecast

CPCB (2000) gives data for 1996-97 on actual coal supplied and estimated coal demand for 2001-02, 2006-07 and 2009-10 (See Table 3.2).

It may be noted that the share of power in the total demand is estimated to increase from 68.5 per cent in 2001-02 to 72.5 per cent in 2009-10.

Table 3.1

Company-wise Coal Production During Last Five Years

(In million tonnes)

Company	1996-1997		1997-1998		1998-1999		1999-2000		2000-2001		2001-2002*	
	Target	*Actual*	*Target*	*Actual*	*Target*	*Actual*	*Target*	*Actual*	*Target*	*Actual*	*Target*	*Actual*
ECL	31.50	29.65	32.50	27.42	32.00	27.16	29.00	25.12	28.00	28.03	20.18	19.18
BCCL	29.50	27.14	30.70	30.92	32.30	27.17	27.50	27.90	29.50	25.97	21.65	18.02
CCL	35.20	32.18	33.50	33.08	35.00	32.17	33.50	32.40	34.00	31.75	24.69	22.30
NCL	37.00	37.01	37.00	37.12	37.00	36.52	37.50	38.43	39.00	41.40	28.82	30.63
WCL	29.50	31.23	30.50	32.52	32.00	31.75	32.00	33.86	33.00	35.20	24.86	26.99
SECL	52.50	55.30	55.50	56.63	58.70	57.56	58.00	58.75	60.00	60.33	44.76	44.71
MCL	36.00	37.36	39.50	42.17	41.00	43.51	41.00	43.56	43.00	44.80	32.04	34.11
NEC	0.80	0.75	0.80	0.69	0.85	0.64	0.60	0.57	0.50	0.66	0.36	0.42
Total CIL	**252.00**	**250.62**	**260.00**	**260.55**	**268.85**	**256.48**	**259.10**	**260.59**	**267.00**	**268.14**	**197.36**	**196.36**
SCCL	30.20	28.73	31.00	28.94	31.00	27.33	31.00	29.56	31.67	30.27	22.74	21.90
Captive Collieries	6.45	6.28	6.45	6.31	6.65	8.46	8.80	9.82	9.40	11.22	8.00	8.98
Grand Total	**288.65**	**285.63**	**297.45**	**295.80**	**306.50**	**292.27**	**298.80**	**299.97**	**308.07**	**309.63**	**228.10**	**227.24**

Note: (i) *: Provisional (ii) Does not include Meghalaya Coal.

Source: Coal India Limited, *Annual Report 2001*, New Delhi, India.

Table 3.2

Coal Demand Forecast

Sector	1996-97 (actual)		2001-02 (estimate)		2006-07 (estimate)		2009-10 (estimate)	
	Million Tonnes	Percent-age	Million Tonnes	Percent-age	Million Tonnes	Percent-age	Million Tonnes	Percent-age
Power	199.00	69.5	285	60.5	400	70.8	500	72.5
Steel	25.53	8.9	49	11.8	60	10.6	68	9.9
Others	61.93	21.6	82	19.7	105	18.6	122	17.7
Total	**286.46**	**100.0**	**416**	**100.0**	**565**	**100.0**	**690**	**100.0**

Source: CPCB, 2000.

Imports and Exports

Coal is under the Open General Licence (OGL) list. So coal can be freely imported by companies subject to import duty. Coking coal is being imported by Steel Authority of India Limited (SAIL) and other steel manufacturers mainly to bridge the gap between their requirements and indigenous availability and to improve the quality of overall blend for technological reasons. Coal based power plants, cement plants, captive-power plants, sponge iron plants, industrial consumers and coal traders are importing non-coking coal on consideration of transport logistics and commercial prudence as well as against export entitlements. Mainly pig-iron manufacturers and iron and steel sector consumers using mini-blast furnace import coke. Due to high (> 17 per cent) ash content in the coal supplied to the steel industry, the import dependency for coking coal is steadily increasing.

Table 3.3

Import of Coal and Coal Products

(in million tonnes)

Coal	1998-99	1999-2000	2000-01	2001-02 (April-Sept. 2001) (Data Provisional)
Coking Coal	10.02	10.99	11.06	4.80
Non-Coking Coal	6.51	8.71	9.87	4.85
Coke	1.57	2.41	2.42	1.09
Total Import	**18.10**	**22.11**	**23.35**	**10.74**

Source: Website of Coal India Limited, *www.coalindia.nic.in.*

The custom duty rates on imported coal during 2001-02 are given in Table 3.4.

Table 3.4

Custom Duty Rates on Imported Coal

Type of Coal		*Basic Duty*	*Other Duties (SAD)*	*Total (Effective) Import Duty*
Coking	Having up to 12 per cent ash	5%	4.2%	9.2%
Coal	Having 12 per cent and more ash	15%	4.6%	19.6%
Coke*		15%	4.6%	19.6%
Non-Coking Coal		25%	5%	30%

Note: * The basic import duty for coke (metallurgical) when imported by a manufacturer of pig-iron or steel using blast-furnace has been fixed at 5 per cent as per amendments contained in the Notification of Government of India in Ministry of Finance (Department of Revenue) *vide* No. 44/2001-Customs dt. 26.4.2001. Effective import duty for this category works out to 9.2 per cent.

India exports coal to the neighbouring countries. The quantum of coal exported by CIL during 2000-01 to the neighbouring countries was 0.043 million tonnes. During 2001-2002 (up to December, 2001) the quantity of coal exported by CIL was 0.016 million tonnes (provisional).

Pricing

On April 1, 2000, the pricing of coal became fully deregulated, *vide* the Colliery Control Order, 2000. Since deregulation, the price of coal with advantageous geographic location (e.g. Rajmahal) is being pegged at a higher level than those facing competition (e.g. Talcher) from imported coal and other fuels.

Royalty on Coal

In terms of Section 9 of the Mines and Minerals (Regulation and Development) Act, 1957, the holder of mining lease shall pay royalty in respect of any mineral removed/consumed. The Central Government can enhance or reduce the rate at which the royalty shall be payable. However, the rates cannot be enhanced more than once in a period of three years. The royalty on coal is considered to be a payment for the depletion of a natural resource but acts as a tax, which is imposed by the Central Government but collected and appropriated by Government of the State where coal production takes place. Prior to the latest revision of royalty rates in October, 1994, the rate was revised in August, 1991, with the rates varying between Rs. 25 per tonne for the lowest grades of coal to Rs. 150 per tonne for the highest grade.[1] The next increase in royalty rates was made w.e.f 11.10.94 (Table 3.5). The enhancement of rates was made, keeping in view the interests of both the consumers and the coal producing states and at the same time keeping in mind the impact on the cost of production and the development of the coal mining industry. The revised rate of royalty has not been extended to the state of West Bengal as that state is levying and collecting cess on coal under the Cess Acts. The revised rates can be extended to the state after the cesses levied by state government are withdrawn or their Cess Acts are struck down by the Court of Law.

Table 3.5

Rates of Royalty on Coal

(Rupees per metric tonne)

	Royalty w.e.f. 1.8.91	*Royalty w.e.f. 11.10.94*
Group–I (Coking Coal SG I, II. WG I)	150	195
Group–II (Coking Coal WG II, III SC I, II Non-coking A, B)	120	135
Group–III (Coking Coal WG IV, Non-coking C)	75	95
Group–IV (Non-coking D, E)	45	70
Group–V (Non-coking F, G)	25	50
Group–VI (Coal produced in Andhra Pradesh)	70	75

Source: Website of Coal India Limited, *www.coalindia.nic.in*.

1. Coal, in India, is classified into seven categories, A-G, based on ash content, gross calorific value and useful heat value. See Table 3.9.

Environmental Problems

Environmental damage and pollution arise during the mining, processing and use of coal in various applications. There are many environmental regulations/guidelines relating to coal mining. Environmental standards, both air quality and water quality, have been prescribed for industries using coal as a fuel. These regulations are in the nature of command and control regulations but there are cost-effective technologies for controlling pollution to achieve the prescribed standards.

Among the environmental problems arising from the use of coal in power generation and industrial applications, the problem of using coal with high ash content has been highlighted in many research studies and government reports. This problem assumes importance because of the increased dependence on coal with high ash content for thermal power generation and many industrial applications which are associated with emissions in the form of particulate matter, toxic elements, fly ash, oxides of nitrogen, sulphur and carbon besides large volume of water for cooling and land requirement for ash disposal. Reduction in ash content is possible *via* coal beneficiation. This will not only reduce the ash content to the required level but also enrich the coal for better thermal efficiency, apart from improving plant availability, reducing operating costs and the load on transport system, and solid waste generation. Even though the Government of India promulgated Gazette Notifications GSR 560(E) dated September 19, 1997 and GSR 378(E) dated June 30, 1998, on the use of beneficiated/blended coal containing ash not more than 34 per cent with effect from June 2001 in power plants located 1000 kms from pit head and in power plants located in critically polluted areas, urban areas and ecologically sensitive areas, the responses to these notifications have been poor. We consider the use of an economic instrument, namely, a cess on coal based on its ash content along with the use of revenue from the cess for coal beneficiation and technological upgradation as an incentive measure.

Inferior Coal and Beneficiated Coal

Indian coal contains 30-40 per cent ash (on an average), which is far higher than in many other countries.[2] Table 3.6 presents examples of typical coals from five areas in India.

2. Over 200 million tonnes of coal reach the consumers with ash content averaging 40 per cent (CPCB, 2000).

Table 3.6

Analyses of Typical Indian Coals from Selected Regions

Rank	*Jharia*		*Uttar Pradesh*	*Renusagar*	*Singrauli*	*Neyveli*
	Medium volatile bituminous	*High volatile bituminous*	*High volatile bituminous*	*Sub-bituminous*	*Sub-bituminous*	*Lignite*
As Received						
Ash, %	38.9	31.6	28.0	28.6	31.5	4.5
Moisture, %	1.1	6.9	10.0	14.9	7.9	53.1
Moisture & Ash Free						
Volatile, %	25.3	37.2	41.0	45.1	47.4	57.1
Carbon, %	83.6			74.1	71.9	70.3
Hydrogen, %	4.5			4.8	5.0	5.2
Oxygen, %	9.9			18.6	20.3	23.1
Nitrogen, %	1.3			1.4	2.0	0.5
Sulphur, %	0.7	1.8	0.8	1.1	0.8	0.9
Lower Heating value, MJ/kg	33.0	30.4	30.7	28.4	27.3	28.4
Hardgrove Grindability, ^{O}H	63	60	50	56	50	95+

Source: Singer (1981).

Coking coal contains less ash than power grade coal. Coal ash has serious environmental implications for air, water and land pollution. Reduction of ash in coal on the one hand reduces the transportation cost and air pollution during the transport of coal, and on the other hand it improves the gross calorific value and hence useful heat value of coal, which is of great significance in cases where coal is used for generation of energy/power. Coal containing less ash yields other benefits such as increased boiler efficiency, reduction in auxiliary energy consumption, decrease in maintenance cost and decrease in coal handling and storage costs.

Most of the coking coal washing plants in India are old and have low unit capacity. Of the 20 washeries currently in operation, only one has an annual output of over 1MT of washed coking coal and three have individual outputs of less than 3 lakh tonnes a year. In addition, capacity utilisation of plants is typically in the order of 50-65 per cent. The CIL is reported to have been facing difficulty in maintaining its contractual obligation of supplying coking coal with 17 per cent maximum ash content to SAIL.

According to (CPCB, 2000), only 20 per cent of the total coal transported to the power plants is of superior grade with ash content 24 per cent or less and the remaining 80 per cent is of inferior grade with ash content ranging from 24 to 45 per cent. In power plants, the ash content of the coal has an effect on coal costs, transportation costs of coal, ash handling, maintenance costs, generation capacity and availability and forced outage rate. Of the total ash (30-40 per cent on an average in thermal grade coal), about 20 per cent is deposited in the form of bottom ash and the remaining 80 per cent in fly ash. For a typical 210 megawatt unit plant, (coal with 30 per cent ash) on an average, 2.69×105 tonnes of ash are generated per year. As a result, the dust concentration in the flue gas in the absence of any control measure would be 37.5 gms/Newton metre (Nm3). Annexure A-3.1 provides the list of power plants, which are required to use beneficiated coal from June 2001. Currently, about 37 per cent of non-coking coal is transported to power plants located beyond 1000 km from the coalfields. Furthermore, under the proposal, there would be a phased introduction of maximum 34 per cent ash in power station feed between 2000 and 2004 with all new generating capacity constructed after June 2000 designed to use this fuel. Despite having long-term targets for the construction of 60 MT annual washing capacity for non-coking coal, coal washing has not kept pace with demand. An attainable and reasonable goal for the washing of Indian raw coal is to reduce the ash content to at least 30-25 per cent (World Bank, 1997).

Sachdev (1992) has estimated the savings due to reduced ash content split into different parts for four power plants using data from four representative Indian units in three power stations and typical coal data. A substantial economic incentive for firing washed coals in these power plants was identified. A break-even cost analysis established the following: (i) A premium of about $ 0.55/tonne could be paid for each percentage point reduction in the ash content of the typical high-ash bituminous coal fired in older, existing power plants; and (ii) cleaning high-ash coals for use in newer plants that were designed for high-ash coals was found somewhat less attractive. A premium of about $ 0.40/tonne could be paid by newer plants for each percentage point reduction. It would be seen from Table 3.7 that the estimated savings arise mainly from reduced maintenance cost within the power plants, increased plant availability, and reduced coal transportation costs.

Table 3.7

Savings Due to Reduced Ash Content, Split into Different Parts for Four Power Plants

	A, old %	*B, newer %*	*C, old %*	*D, old %*
Fuel (free on board)	2	6	2	4
Transportation	49	27	19	68
Operation	0	0	11	0
Maintenance	39	27	14	23
Derate*	0	22	34	0
Availability	10	18	20	5
Total	**100**	**100**	**100**	**100**

Note: * For example, high ash content may result in restricted mill throughput and higher energy consumption in mills.

Source: Sachdev (1992).

According to the Central Fuel Research Institute, Dhanbad, if the entire power coals which are to be transported over 1000 km in the year 2001-02 are washed and the ash level is brought down from 40 to 34 per cent, it is likely to reduce 11 million tonnes in transport (reduction in tonnage to be transported), 8 million tonnes of fly ash, and 23 million tonnes of carbon dioxide emissions. According to the CPCB, in the majority of cases, it is technically feasible and economically viable to reduce the ash content of raw coal to 32-34 per cent, with net benefits to power plants.

Reduction in ash content in coal can be achieved either by selective mining or by beneficiation of coal. Beneficiation is the most frequently considered option for this. However, washing of Indian coal is also perceived to be generally difficult as it is rich in minerals and thus contains near gravity impurities, making it hard to separate the ash from coal by gravity separation. According to an estimate for deep washing (which reduces ash content by 11 per cent) the system cost for a million tonnes of coal is Rs. 250-400 million. There are, of course, additional costs of land, water and managing washery rejects (TERI, 2001). According to another study by a German consultant (CPCB, 2000), the cost of washing coal ranges from Rs. 103/tonne to Rs. 172/tonne for reducing the ash level to 34 per cent, the average being Rs 132/tonne. A recent study carried out by Central Mine Planning and Design Institute for three coal beneficiation plants gives the cost estimates for reducing the ash content of coal from 40 per cent to various levels of ash (Table 3.8). It would be seen from Table 3.8 that the marginal beneficiation cost increases at an increasing rate beyond the reduction of ash below 30 per cent.

Table 3.8

*Cost of Beneficiation of Coal**

(Rs. per Million Kcal)

Sl. No.	*Ash Content*	*Ash Reduction to*			
		34%	*32%*	*30%*	*25%*
	38%	**I. Dipika Mine**			
1. ROM (Run of Mine) Cost	74.53				
2. Beneficiated Coal Cost		95.38	102.84	116.73	151.92
3. Beneficiation Cost (2-1)		20.85	28.31	42.20	77.39
4. GCV = Kcal/kg	4166	4397	4585	4773	5244
	41%	**II. Kalinger Mine**			
1. ROM (Run of Mine) Cost	81.12				
2. Beneficiated Coal Cost		112.69	116.71	127.96	171.97
3. Beneficiation Cost (2-1)		31.57	33.59	46.84	90.85
4. GCV = Kcal/kg	4166	4337	4227	4715	5186
	42%	**III. Piparwar Mine**			
1. ROM (Run of Mine) Cost	63.92				
2. Beneficiated Coal Cost		79.89	80.11	80.74	100.56
3. Beneficiation Cost (2-1)		15.97	16.91	16.82	36.64
4. GCV = Kcal/kg	3700	4410	4598	4786	5256

Note: * Based on the study conducted by Central Mine Planning and Design Institute reported in Sankar and Mathur (1998).

Singh and Bora (2000) have shown that selective mining has great potential in improving quality of coal in the Talcher mines of HCL. It is techno-economically feasible, in most cases, to selectively mine the coal, without the impurities. The cost of a surface miner is around Rs. 50 million. The total system cost to produce one million tonnes of coal per year is estimated at around Rs. 120 million and this could reduce ash content by about 13 per cent.

Coal Cess Based on Cost of Beneficiation[3]

Since Indian coal is high in ash content and as the use of coal with high ash content imposes costs on society, an ecocess on coal differentiated by ash content is proposed. This cess would induce efforts to bring down the ash content. In designing the ecocess the current system of grading of coal has been used. In India, while coking coal is graded on the basis of ash and moisture content in coal, non-coking coal is graded on the basis of useful heat value (UHV) of coal.[4] To make the grading of coking and non-coking coal comparable, UHV in the case of non-coking coal has been converted into ash and moisture content. According to the 'polluter pays' principle, the environmental tax on a polluting activity/good should be set equal to the marginal damage caused by that activity/good. In the absence of reliable information about the damage caused at the margin due to ash in coal, an alternative would be to consider the cost of reducing the ash in coal to the required extent. Using the existing estimates of cost of ash reduction in coal the rates of cess on coal have been estimated.

Since categorisation of coal by the Coal Board for determining royalty and others is done on the basis of only the ash content of coal, UHV of non-coking coal is converted into ash content by assuming a six per cent moisture content in coal on an average. The estimated ash in various grades of non-coking coal is as follows.

As discussed earlier, three types of estimates of cost of beneficiation of coal are available. While one estimate is for 11 per cent reduction in ash per tonne of coal, the second estimate is for reduction in ash up to 34 per cent. The third gives the cost estimates for reducing the ash content of coal from 40 per cent to various levels of ash per million kcal for three different mines. This study

3. OECD countries have VAT or/and non-VAT taxes on coal. See Annexure A-3.2. Only a few countries have environmental taxes based on atmospheric emissions or carbon content of coal or perceived health damage. Carbon tax is becoming popular in European countries because of their concern about greenhouse gas emissions and their commitment to Kyoto Protocol. India, as a developing country, has no commitment to reduction of greenhouse gases. Indian environmental legislations do not provide for taxes on emissions. The coal industry argues that coal is not a manufactured product. Hence, we consider a cess on coal based on its ash content.

4. The UHV in kcal/kg is defined by the following formula: UHV = [8900 – 138 × (per cent ash + per cent moisture)] (kcal/kg).

has used the third estimate for computing the cost of beneficiation per tonne of coal for reducing ash from 40 per cent to various lower levels (Table 3.10).

Table 3.9

Calculation of Ash for Non-Coking Coal

(with 6 per cent moisture content)

Grade	*UHV kcal/kg*	*Ash (per cent)*
A	- > 6200	- > 13.56
B	5600-6200	13.56-17.91
C	4940-5600	17.91-22.69
D	4200-7940	22.69-28.06
E	3360-4200	28.06-34.14
F	2400-3360	34.14-41.10
G	1300-2400	41.10-49.07

Table 3.10

Cost of Beneficiation of Coal

(Rs. per tonne of coal)

S. No.	*Mines*	*Ash Content*	*Ash Reduction to*			
			34%	*32%*	*30%*	*25%*
1.	Dipika Mine Beneficiation Cost per tonne of Coal	38%	91.68	129.8	201.42	405.84
2.	Kalinger Mine Beneficiation Cost/tonne of Coal	41%	136.92	161.11	220.85	471.15
3.	Piparwar Mine Beneficiation Cost/tonne of Coal	42%	70.43	74.44	79.31	192.58

Source: Calculated from Table 3.8.

The cost of beneficiation is given in Rs. per million k/cal for various levels of ash reduction in Table 3.8. On dividing this by the corresponding GCV available in k/cal per kg we get the quantity of coal in kg. From this the cost of beneficiation per tonne of coal has been computed. Estimates of the cost of beneficiation of coal are presented in Table 3.10. It would be seen from Table 3.10 that for reducing the ash content in coal to 34 per cent, the average cost of beneficiation for one percentage point reduction in ash ranged between Rs. 8.80 to Rs. 23 per tonne of coal. The cost of beneficiation for one percentage point reduction in ash, on an average, from 34 per cent to 30 per cent ranged between Rs. 2.21 to Rs. 27.43 per tonne of coal. Further reduction in ash content of coal is estimated to cost between Rs. 23 to Rs. 50.13 for each percentage point reduction in ash (on an average). That is, the marginal beneficiation cost increases at an increasing rate beyond the reduction of ash below 30 per cent. In this study, the optimal level of beneficiation of non-coking coal is taken to be the reduction in ash content of coal up to the level of 34 per cent. This would also be consistent with the current policy of the government.

In the case of coking coal since quality requirements in terms of ash content are higher *vis-à-vis* non-coking coal, the optimal level of beneficiation was taken to be the reduction in ash content of coal up to the level of 18 per cent. On this basis, an ecocess is proposed on coking coal containing ash greater than 18 per cent. Using the estimates of the cost of beneficiation of coal for various levels of ash reduction, the following rates of ecocess on coal were initially proposed (Table 3.11).

Table 3.11

Rates of Ecocess on Coal Based on Beneficiation Costs

	Grade of Coal	*Rate of Ecocess (Rs. per tonne)*	*Ash Per Cent*
A.	**Coking Coal**		
1	S - I	-	< 15
2	S - II	-	15 - 18
3	W - I	40	18 - 21
4	W - II	40	21 - 24
5	W - III	80	24 - 28
6	W - IV	100	28 - 35
7	SC - I	-	< 19
8	SC - II	40	19 - 24
B.	**Non-coking Coal**		
	A	-	< 14
	B	-	14 - 18
	C	-	18 - 23
	D	-	23 - 28
	E	100	28 - 34
	F	125	34 - 41
	G	125	41 - 49

To estimate the burden of proposed ecocess on coal as per cent of coal prices, ideally prices of various grades of coal at the pit-head of various mines would be required. However, this information is available for only a select mines for ROM non-long-flame (NLF) non-coking coal. We therefore calculate the burden of ecocess on coal as per cent of the basic price of ROM (NLF) non-coking coal at the pit-head. It may be noted that the basic price of coal (ROM) for different grades of coal varies across mines. For instance, the basic price of ROM coal is highest among CIL mines at Rs. 1628 of ECL/SP mines and lowest at Rs. 912 of MCL mines. Similarly, for grade G the highest price is at Rs. 479 of ECL/Rajmahal and lowest at Rs. 250 of MCL mines. Further, the final prices of coal can be higher by 7–18 per cent of the basic ROM coal depending upon the quality e.g. premium or long flame (CIL, 2001).

Table 3.12

*Ecocess Based on Beneficiation Costs as Per Cent of Basic Price of ROM (NLF) Non-Coking Coal**

(CIL prices as on 17.8.2002)

Grade of Coal	*Price of Coal (Rs.)*		*Ecocess as Per Cent of*		*Ecocess as Per of Final Price of Coal***	
	(Highest)	*(Lowest)*	*Coal Price (Highest)*	*Coal Price (Highest)*	*(Highest)*	*(Lowest)*
A	1628	912	-	-	-	-
B	1447	819	-	-	-	-
C	1211	674	-	-	-	-
D	974	566	-	-	-	-
E	743	445	13.45	22.47	12.01	20.06
F	620	351	20.61	35.6	18.01	31.79
G	479	250	26.09	50.0	23.30	44.64

Note: * For all grades of coal, price of steam coal and slack coal are price of ROM coal plus Rs. 150 and Rs. 10 respectively.

** Final price of coal is taken to be 12 per cent higher than the price of ROM coal.

For example, for CCL premium collieries the prices of grades A to G were fixed at price of ROM coal + Rs. 164, 154, 144, 136, 44, 34 and 25 respectively. For NCL long flame coal final prices were; price of ROM coal + Rs. 75, 75, 75 and 76 for grades A, B, C and D respectively. Given this the actual burden of the proposed cess would be lower than shown in Table 3.12. For instance, the lowest final price of E grade coal would be Rs. 512 instead of Rs. 445 and thus the proposed ecocess would be around 19 per cent instead of 22.47 per cent shown in Table 3.12. Similarly, the burden of cess as per cent of coal price would be lower for F and G categories also. From the last two columns of Table 3.12 it would be seen that the burden of ecocess as per cent of highest price is in the range of 12-23 per cent and as per cent of lowest price is in the range of 20-45 per cent.

Price of various grades of coking coal (ROM) is presented in Table 3.13. The estimates of burden of ecocess as per cent of ROM coking coal are presented in Table 3.14.

Table 3.13

Selling Price of ROM Coking Coal

(CIL prices as on 31.1.2001)

	SI	*SII*	*WI*	*WII*	*WIII*	*WIV*	*SCI*	*SCII*
Bharat Coking Coal Ltd. (For some collieries)	1914	1598	1385	1147	848	789	1335	1106
Bharat Coking Coal Ltd.	1695	1416	1227	1016	751	699	1183	979
Eastern Coal Fields Ltd.	-	-	1312	1087	803	747	1360	1126
Central Coal Fields Ltd.	-	-	1287	1067	789	734	1241	1027
South Eastern Coal Fields Ltd.	-	-	1096	907	671	625	-	-
Western Coal Fields Ltd.	-	-	1075	890	803	659	-	-
All Subsidiaries	-	-	-	-	-	-	1096	907

It would be seen from Table 3.14 that the burden of ecocess for various grades of coking coal ranges between 2.9 to 16 per cent of the price of ROM coking coal. In terms of the percentage the burden of cess would be less on the final price of coking coal (steam and slack grade).

Feedback on Ecocess Based on Beneficiation Costs

There was opposition to the levy of coal cess from the Ministry of Coal, Central Electricity Authority and the National Thermal Power Corporation. A representative of the Coal Ministry argued that with reduction in customs duties on imported coal, domestic coal faced competition from imported coal. He said that the Kelkar Committee

Table 3.14

Ecocess Based on Beneficiation Costs as Per Cent of Price of ROM Coking Coal

(CIL prices as on 31.1.2001)

Grade of Coal	*Prices (Rs.)*		*Ecocess as Per Cent of Coal Price*	
	Highest	*Lowest*	*Coal Price (Highest)*	*Coal Price (Lowest)*
SI	1695	1914	-	-
SII	1598	1416	-	-
WI	1385	1075	2.9	3.7
WII	1147	890	3.5	4.5
WIII	848	671	9.43	11.92
WIV	789	625	12.67	16.0
SCI	1360	1183	-	-
SCII	1126	907	3.55	4.4

proposal to bring down the duty on non-coking coal from 25 per cent to 5 per cent in the next five years would decrease the competitiveness of domestic coal. The CEA contented that levying of cess on grades E to G of non-coking coal of the order 25 per cent to 50 per cent of the cost of Run of Mine Coal would increase the cost of power generated. NTPC argued that the proposed cess would increase the cost of electricity between 10 to 15 paise per kWh and increase demand for imported coal by power plants located 1050 or more kms from pit heads. Dr. Sihag from TERI argued that coal cess would have an inflationary impact. He said that coal washing is only one of the environmental problems in thermal power plant.

It may be noted that in the responses received from the above organisations, there is no reference to the environmental cost imposed on the society by the use of coal with high ash content. This cost is substantial and an attempt is to be made to reduce it and / or to internalise it. It is true that the imposition of a cess on coal, based on its ash content, would raise the input cost of coal; but there would be associated effects which would result in benefits and reduction in costs in other ways. When an inferior quality of non-coking coal (with high ash content) is used for power generation, the electric utility has to pay higher transportation costs of coal from the mines to the power plants (because the ash is also carried). Besides there is a cost disadvantage arising from reduced plant availability, higher costs of compliance with emission standards and fly ash disposal requirements. To the extent that these various costs are reduced because of the reduction in the ash content, the cost of production of electricity will rise only by a fraction of ecotax imposed. To the extent that there is a rise in the cost of production because of the ecotax that reflects part of the hidden cost that is borne by the society because of the high ash content. The ecotax serves to give incentive for reducing that cost and to internalise it to the extent it continues to exist. CPCB (2000) notes that the net benefits of coal beneficiation are positive. A coal cess is an instrument for internalising the negative externality.[5]

It is likely that imposition of the coal cess would encourage power plants located 1000 km from pit heads to import coal with lower ash content. This may be considered a beneficial effect from the environmental point of view. Further, with a foreign exchange reserve of over US$ 120 billion, the foreign exchange constraint is not binding. However, we recognise the concerns about inflation in the short run and the need for giving time to increase the capacity of plants engaged in coal washing. Hence, we propose cess rates lower than the beneficiation costs, at least initially. An advantage of the coal cess is that revenue from the cess can be earmarked for investments in coal beneficiation and technological upgradation schemes in the coal sector. We wish to reiterate that a cess on coal is one of the instruments for solving the environmental problems in the coal sector. In fact, our proposal is complementary to the existing standards and control regime relating to emissions and effluents and regulations regarding coal mining.

Proposed Rates of Ecocess on Coal

We propose lower rates than recommended earlier (about 50 per cent of the costs of beneficiation) to allow adjustment time to the coal miners, coal washeries and the power plants and other

5. The saving in the cost of transportation of coal will counteract the increase in cost caused by the cess.

users. Raising the rates of ecocess to the coal beneficiation costs may be considered in about three years from the date of implementation of ecocess on coal. The proposed rates for now are given in Table 3.15.

Table 3.15

Proposed Rates of Ecocess on Coal

	Grade of Coal	*Rate of Ecocess (Rs. per tonne)*	*Ash Per Cent*
A.	**Coking Coal**		
1	S - I	-	< 15
2	S - II	-	15 - 18
3	W - I	20	18 - 21
4	W - II	20	21 - 24
5	W - III	40	24 - 28
6	W - IV	50	28 - 35
7	SC - I	-	< 19
8	SC - II	20	19 - 24
B.	**Non-coking Coal**		
	A	-	< 14
	B	-	14 - 18
	C	-	18 - 23
	D	-	23 - 28
	E	50	28 - 34
	F	70	34 - 41
	G	70	41 - 49

The burden of cess as per cent of coal price is given in Table 3.16. From the last two columns of the table it would be seen that the burden of ecocess as per cent of the highest price is in the range of 6–13 per cent and as per cent of the lowest price is in the range of 10–25 per cent.

For example, for Central Coalfields Limited (CCL) premium collieries, the prices of grades A to G were fixed at price of ROM coal + Rs. 164, 154, 144, 136, 44, 34 and 25, respectively. For Northern Coalfields Limited (NCL) long flame coal final prices were: price of ROM coal + Rs. 75, 75, 75 and 76 for grades A, B, C and

Table 3.16

*Ecocess as Per Cent of Basic Price of ROM (NLF) Non-Coking Coal**

(CIL prices as on 17.8.2002)

Grade of Coal	*Price of Coal (Rs.)*		*Ecocess as Per Cent of*		*Ecocess as Per Cent of Final Price of Coal***	
	Highest	*Lowest*	*Coal Price Highest*	*Coal Price Lowest*	*Highest*	Lowest
A	1628	912	-	-	-	-
B	1447	819	-	-	-	-
C	1211	674	-	-	-	-
D	974	566	-	-	-	-
E	743	445	6.7	11.2	6.0	10.03
F	620	351	11.29	19.94	10.08	17.8
G	479	250	14.61	28.0	13.07	25.00

Note: * For all grades of coal, prices of steam coal and slack coal are prices of ROM coal plus Rs. 150 and Rs. 10, respectively.

** Final price of coal is taken to be 12 per cent higher than the price of ROM coal.

D, respectively. Given this the actual burden of the proposed cess would be lower than shown in Table 3.16. For instance, the lowest final price of E grade coal would be Rs. 498 instead of Rs. 445 and thus the proposed ecocess would be around 10 per cent instead of 11.2 per cent shown in Table 3.16. Similarly, the burden of cess as per cent of coal price would be lower for F and G categories also. From the last two columns of Table 3.16 it would be seen that the burden of ecocess as per cent of the highest price is in the range of 6-13 per cent and as per cent of the lowest price is in the range of 10-25 per cent.

The prices of various grades of coking coal (ROM) are presented in Table 3.13. The estimates of burden of ecocess as per cent of ROM coking coal are presented in Table 3.17.

It would be seen from Table 3.17 that the burden of ecocess for various grades of coking coal ranges between 1.44 to 8 per cent of the price of ROM coking coal. In terms of percentage the burden of cess would be less on the final price of coking coal (steam and slack grade).

Table 3.17

Ecocess as Per Cent of Price of ROM Coking Coal

(CIL prices as on 31.1.2001)

Grade of Coal	*Prices (Rs.)*		*Ecocess as Per Cent of Coal Price*	
	Highest	*Lowest*	*Coal Price (Highest)*	*Coal Price (Lowest)*
SI	1695	1914	-	-
SII	1598	1416	-	-
WI	1385	1075	1.44	1.85
WII	1147	890	1.75	2.25
WIII	848	671	4.71	5.96
WIV	789	625	6.33	8.0
SCI	1360	1183	-	-
SCII	1126	907	1.77	2.2

Estimate of Revenue from Ecocess

The impact of the ecocess on revenue would depend on factors influencing the supply of and demand for beneficiated coal. It may take one or two years to increase the capacity of existing coal washeries or commission new plants for coal beneficiation. Thermal power plants and coal-using industries which are located 1000 km or more from the pit heads and near the coastal areas would make an assessment of the comparative costs of using coal with ash content of 34 per cent or more, beneficiated coal, and imported coal and then choose the least cost option. The demand for coal would depend largely on its own-price elasticity of demand and the output growth rates of coal using activities.

If the own price elasticity of demand is zero and it takes more than a year to increase the supply of beneficiated coal, the estimated revenue generation from ecocess would be about Rs.176 crore from coking coal and Rs. 835 crore from non-coking coal. Our conjecture is that the revenue realisation would be around Rs.100 crore from coking coal and about Rs. 500 crore from non-coking coal a year after the imposition of the cess because of the import option and the price elasticity being different from zero (between -0.1 and -0.2). We want to emphasize that the purpose of the coal cess is not to generate revenue to the government but to signal the users about the environmental costs of using coal with high ash content. We recommend the use of cess revenue to support investments in coal beneficiation and adoption of environmentally sustainable technologies and practices in the coal sector so that the costs of adopting environment-friendly inputs and technologies can be lowered.

Clean Coal Fund

Revenues generated from ecocess on coal may be used to set up Clean Coal Fund which could be utilised for setting up infrastructure for coal washing, selective mining, research and development to identify activities for gainful utilisation of coal ash and safe storage and disposal of the residual ash. Higher and more consistent coal quality will increase generation efficiency and plant availability. Energy conservation from reduced transportation load and reduction in CO_2 emissions from 0.983 to 0.866 kg per kWh will be additional benefits.

Though an ecocess on lower quality of coal will incentivise the demand for and supply of good quality coal, this effect needs to be strengthened further by bringing in reforms in the power sector which would create incentives for sound financial and environmental performance of power plants.

The clean coal fund jointly managed by the principal stakeholders, the coal companies, the power plants and the MoEF/CPCB should work towards ensuring compliance with environmental regulations. For instance at present, power plants beyond 1000 km from the coal field, as also those in certain urban areas/ecologically sensitive areas have to utilise coal with less than 34 per cent ash content. In effect, this means that washeries have to be set up to ensure that these plants are supplied the coal of the designated quality. Here, there is an ambiguity in the responsibility of the coal producer and the electricity producer. There is no agreement as to who has the responsibility to implement this notification. Both have their own difficulties in establishing these washeries. The electricity producers have argued that as the washery necessarily needs to be set up near the coalfield, from the point of view of administrative issues and logistics, it makes sense for the coalfield to establish a washery and charge a premium for washed coal. Coal producers have, however, argued that given their financial position, they are not capable of establishing washeries. The Clean Coal Fund can take up this and other such issues to be sorted out.

Conclusion

Coal containing high ash content causes serious environmental pollution and health hazards both during transportation and handling, and use in industrial applications and power generation. There is evidence to show that coal beneficiation reduces the ash content in the coal thereby improving thermal efficiency, plan availability and plant output, and reducing operation and transport costs of thermal power plants. Studies show that the use of coal with 32-34 per cent ash content is likely to result in net benefits to power plants. With a view to incentivise the demand for and supply of coal containing low ash content an ecocess on coal differentiated by ash content is proposed. We feel that this needs to be complemented with reforms in the power sector, which would create incentives for sound financial and environmental performance of power plants.

Revenues generated from ecocess may be used to set up Clean Coal Fund which could be utilised for setting up infrastructure for coal washing, selective mining and research and development to identify activities for gainful utilisation of coal ash and safe storage and disposal of the residual ash.

Annexure A-3.1

Thermal Power Plants Required to Use Beneficiated Coal from June 2001

A: Existing

S. No.	*Name of Thermal Power Station*	*Capacity (MW)*	*Category*	*Estimated Annual Coal Requirment for 1000-2000(MTPA)*
1.	Badarpur	705	UA	2.75
2.	Indraprastha	278	UA	0.87
3.	Rajghat	135	UA	0.58
4.	Faridabad	165	UA	0.80*
5.	Panipat (Units 1-5)	650	>1000 km	3.60*
6.	Bhatinda (Units 1-4)	440	>1000 km	1.98
7.	Ropar (Units 1-6)	1260	>1000 km	5.08
8.	NCR Dadri	840	>1000 km	4.00
9.	Harduaganj	425	>1000 km	1.06
10.	Panki	274	UA	0.79
11.	Paricha	220	>1000 km	0.89
12.	Kota (Units 1-5)	850	UA	3.65
13.	Sabarmati	410	UA	1.32*
14.	Wanakbori (Units 1-6)	1260	>1000 km	6.06
15.	Gandhinagar	660	UA	3.00*
16.	Ukai	850	>1000 km	3.36*
17.	Sikka (Units 1-2)	240	>1000 km	1.00*
18.	Bhusawal	478	>1000 km	2.24
19.	Koradi	1080	UA	5.50*
20.	Nashik	910	>1000 km	3.60
21.	Trombay	1150	UA	Oil/Coal
22.	Dahanu	500	SA	2.01
23.	CESC & Southern	190	UA	0.72*
24.	Tatagarh	240	UA	0.96*
25.	DPL	390	C P A	0.49
26.	Muddanur	420	>1000 km	2.37
27.	(Rayalaseema)	630	UA	2.97
28.	North (Chennai -1)	450	>1000 km	1.92*
29.	Ennore	840	>1000 km	4.38
30.	Raichur (1-4)	840	>1000 km	4.39
31.	Mettur	1050	>1000 km	4.08*
32.	Tuticorin (1-5)	820	CPA	1.84
33.	Bokaro Durgapur	350	CPA	1.00
	Sub-Total (A)	**20,000**		**79.05**

B: Ninth Plan

S. No.	Name of Thermal Power Station	Capacity (MW)	Category	Estimated Annual Coal Requirment for 1000-2000(MTPA)
34.	Bhatinda 5&6	420	>1000 km	1.88*
35.	Wanakbon-7	210	>1000 km	1.00*
36.	Gandhinagar-7	210	>1000 km	0.95
37.	Raichur (5-6)	420	>1000 km	2.14
38.	North Chennai II	1050	UA	IC
39.	Mangalore	1000	>1000 km (Imported coal)	IC
40.	Tranagallu	260	>1000 km	IC
41.	Suratgarh-1	500	>1000 km	IC
	Sub-Total (B)	**4070**		**5.97**

Note: Total coal consumption based on 1999-2000 data up to Ninth Plan 87.14 MTPA.

*: Revised based on data provided by SEBs/utilities.

IC: Imported Coal.

UA: Urban Agglomerates.

CPA: Central Plan Assistance.

Source: Central Electricity Authority.

Annexure A-3.2

Taxes on Coal in OECD Countries

Taxation of energy products vary widely between individual energy products and across OECD member countries. Only rarely has an attempt been made to tax all sources of energy on a consistent basis and very large differences in tax rates for different types of energy reflect domestic political pressures (low or negative rates of taxation in some) or the desire to reduce the tax burden on fuels predominantly used as industrial inputs.

Coal is the least taxed of all primary energy sources. Most coal taxes consist of a VAT levied on household use (Austria, Belgium, France, Germany, Ireland, Italy, the Netherlands, New Zealand, Spain, Sweden and the United Kingdom). Australia and United States are the only countries that do not have a VAT, general consumption tax or turnover tax. In addition, other kinds of special taxes are levied in different countries.

In Denmark an excise tax, completely refundable to the industry, is imposed on household use. This tax was raised under 1994 tax reform, to be fully implemented in 1998. The environment tax is in the form of a CO_2 tax on energy consumption based on the CO_2 content of each fuel at combustion. The carbon tax on coal is only 50 per cent refundable to the industry.

Finland introduced Europe's first explicit carbon tax, imposed on fossil fuels according to their carbon content and incorporated as a surtax into the excise duty on fossil fuels. Initially set at a low rate of Mk 24.5 per tonne of carbon, it was doubled to Mk 50 in 1993.The excise duty on fossil fuels was restructured upward in January 1994 and then again in 1995. The tax is split into a fiscal component and a carbon/energy component (which replaced the pure carbon component) with relative weights of 60/40 in terms of total revenues. Exemptions or reduced rates are not applied for industries. However, products used as raw materials in industrial production are exempted from the tax.

Introduced in 1988, the environmental charge on fuels in the Netherlands, replaced a system of programme specific levies to finance environmental policy expenditures. Fuel was chosen as base as it provides a 'general' link with the 'polluter pays' principle and is a rough indicator for economic activities resulting in pollution. In 1992, the entire tax base was changed to one based on 50/50 on energy/carbon content. There are no excise duties on coal here.

In Norway, the current tax system for fossil fuels already include taxes on atmospheric emissions of CO_2, SO_2 and lead. From 1 July 1992, a carbon tax was introduced for certain coal and coke applications.

In 1990, Sweden introduced a carbon tax as part of a systematic reform of energy taxation to take into account of various forms of polluting emissions. A carbon tax amounting to SKr 0.25per kg of carbon dioxide was introduced on 1 January, 1992 and levied at a rate of SKr 620 per tonne on coal. At the beginning of 1993, the energy tax on fuels was abolished for manufacturing and commercial horticulture. CO_2 taxes were raised from SKr 0.25 per kg to SKr 0.32 per kg of carbon dioxide for non industrial consumers. A sulphur tax is levied on sulphur content in coal in Sweden at the rate of SKr 30,000 per tonne of sulphur.

The United States imposes a black lung disability tax of US $ 1.10 per tonne of coal (except lignite) from underground mines and US $ 0.55 per tonne of coal from surface mines. Rate is capped at 4.4 per cent of the price of coal when sold by the consumer. Some state and local governments impose severance taxes on coal.

Annexure Table A-3.2.1

Taxes on Coal in OECD Countries

Countries	*Rates of VAT or General Consumption Taxes on Coal (residential use only) in OECD Countries in 1994*	*Non VAT Taxes on Coal*
Australia		NA
Austria	20	0
Belgium	12	0
Canada	7	0
Denmark	0	Excise Tax[a] DKr 690/t, Environment Tax[a] DKr 242/t
Finland	22	Additional Tax (carbon/energy): a Mk 67.2/t
France	18.6	0[a]
Germany	15	0[a]
Greece	0	-
Iceland	0	0
Ireland	12.5	0[a]
Italy	9	0[a]
Japan	3	0
Luxembourg	0	0[a]
Mexico	10	-
The Netherlands	17.5	Environment Tax Gld 22.64/tonne
New Zealand	12.5	-
Portugal	16	0
Spain	15	0[a]
Sweden (1993)	25	Carbon Dioxide Tax SKr 0.208/kg (SKr 0.208/kg), General Energy Tax[a] (SKr 0.239/kg), Sulphur Tax: SKr 30/kg
Switzerland	0	Special tax for emergency fund[a] N/A
Turkey	15	-
United Kingdom	8	-
United States		Black lung disability tax of US $ 1.10 per tonne of coal (except lignite) from underground mines and US $ 0.55 per tonne of coal from surface mines. Rate is capped at 4.4 per cent of the price of coal when sold by consumer. Some state and local governments impose severance taxes on coal.

Note: a: Rates of tax for industry and households are same.

Source: OECD, 1995 and 2002.

Annexure A-3.3

Earmarking of Revenues

Earmarked revenues from environmental taxes or charges are increasingly viewed as an important mechanism for financing environmental projects in many parts of the world. In most countries, pollution charges and waste discharge or treatment fees are becoming more closely aligned with the cost of providing basic environmental services, and these revenues are increasingly being retained by the bodies that are responsible for financing those services. At a political level, earmarking is gaining support because it is found that such a 'cost-recovery' approach may make it easier to build consensus, remove barriers, and guarantee budget resources to finance environmental programmes (Huber *et al.*, 1998).

Although earmarking of revenues is normally avoided in public finance, there are certain conditions under which it can be justified. These circumstances usually arise when: (a) transactions and administrative costs of environmental programmes are very high, (b) revenues are related to the explicit provision of a service for a fee, or (c) collection of revenues is politically difficult without clear indication of how the moneys will be used. Earmarked funds provide a means of financing environmental improvements when market, policy and institutional failures impede the emergence of the financing mechanisms characteristic of more developed market economies. In transition and developing economies, these problems can include severe financial constraints on enterprises and households, poorly developed banking systems and capital markets, weak or ineffective enforcement of environmental policy, and uncertainties in fiscal systems. It is sometimes argued that in situations where public budgets are volatile and public investments often cut from year to year, due to the need of macroeconomic stabilisation, earmarked funds can provide secure and sustained financing for environmental programmes. Even in countries that are relatively more stable and developed, such as Poland or the Czech Republic, public and private sector resources for the environment are constrained.

Earmarked 'Environmental Funds' are being used extensively in transition economies, to address broad environmental objectives including education and awareness building, research and institution building, and public and private pollution abatement.

In Poland for example, environmental funds accounted for 33 per cent of total environmental conservation expenditure, whereas the comparable share for the funds in Hungary, Slovenia and Lithuania was about 20 per cent. In

Russian federation, however, funds provided less than five per cent of finance for environmental projects.

Based on extensive reviews of existing funds (OECD, 1995; Lovei, 1999; Huber *et al.*, 1998) it is now generally accepted that financing through earmarking of funds are effective only if the underlying reasons for the environmental problems are simultaneously tackled at the policy level. Although incentive effects are important by-products of earmarked direct environmental taxes, efforts to strengthen enforcement capabilities of environmental agencies and upgrading information systems are pre-eminent in successful funds. Without strengthened environmental regulations and enforcement, environmental funds may end up contributing to existing distortions. Also, the benefits of earmarking are found to be most pronounced when environmental revenues are earmarked to decentralised programmes. To date, experience in Latin America and the Caribbean has demonstrated that earmarking programmes are most successful where: (a) taxes or incentives are linked to existing collection mechanisms, and (b) revenues are made available to decentralised authorities for environmental programming or for institutional strengthening (Huber *et al.*, 1998). Earmarked funds have also been found to be effective when they also strengthen the self-financing capacity of beneficiaries and tackle the causes of financial constraints. Thus co-financing emerges as an important requirement for disbursement of the funds.

Finally, transparency and accountability in the operation of earmarked funds are essential to avoid *ad hoc* political influence and mismanagement of public funds. Mechanisms for stakeholder participation in the decision making can go a long way in improving the transparency and efficiency of the operations. For example, both in Poland and Czech Republic, local authorities play very important roles in the establishment of priorities and in the selection of projects to be funded. The priorities identified by the local authorities thus serves as a basis for decision making at the national level (Francis and Petkova *et al.*, 1999).

Earmarking of environmental funds to overcome or mitigate numerous challenges has come a long way in the last decade or so. In many Latin American countries, the funds have accelerated environmental improvements, leveraged additional finance for environmental investments, and strengthened domestic capacity for project preparation. Experience so far shows that the evolution, effectiveness, and potential of funds is closely linked with the broader progress associated with economic and political reforms, as well as with developments in national environmental policy frameworks.

4

Automobiles

Introduction

Air quality is a crucial factor in determining public health. Some important air pollutants affecting human health are lead, suspended particulate matter (SPM), carbon monoxide (CO), sulphur dioxide (SO_2), nitrogen oxide (NO_x) and toxic substances. Lead affects circulatory, reproductive, nervous and renal systems. Fine particles of SPM may cause lung cancer. CO affects foetal growth in pregnant women, impairs perception and thinking, slows reflexes and can cause unconciousness. SO_2 exacerbates asthma, bronchitis and emphysema. Toxic substances can cause cancer, reproductive problems and birth defects; benzene is a known carcinogen. Air pollution also causes damage to manmade structures and ecology.

Regular monitoring of a few air pollutants is being carried out under the National Ambient Air Quality Monitoring System (NAAQMS). According to the Central Pollution Control Board (CPCB) the level of air pollution in many metro cities exceeds the prescribed standards. As for SO_2, the actual levels exceeded the maximum standard of 80.0 micrograms/m^3 in 7 out of 19 metro cities. In the case of NO_x, 6 of the 19 metro cities had NO_x level exceeding 80 micrograms/m^3. The problem of SPM is more serious. As against the maximum standard of 360 micrograms/m^3, in 13 of the 19 cities the maximum actuals exceeded the maximum standards (in 6 six cities by more than 100 per cent) and in two the actual maximums are closer to the maximum prescribed level (See *Economic Survey* 1998-99, p.151). CPCB data on ambient noise levels in cities reveal that the noise levels in most of the big cities exceed the prescribed standards.

Our knowledge about the causal links between concentrations of air pollutants in the atmosphere and their impact on human health, property values and ecology is poor. There are a few attempts in measuring the health costs of air pollution but these studies are based largely on the methodologies developed and the parametric values derived from studies conducted in USA. See for example, Brandon and Homman (1995) and Sengupta *et al.* (2002).

The ambient air quality at any point of time depends not only on the current (and in some cases past emission levels) but also on meteorological factors. Emissions into air arise from different sources, namely, manufacturing, construction, thermal power generation, transport, household activities, and so on. Some of the sources are point-sources while others are non-point sources. We need a good information system consisting of an emission inventory, links between emissions and air quality, and air quality and health and other damages to estimate statistically the damage value for each pollutant at margin.

We have a legal and administrative system to prevent and control pollution from point sources. This pollution control regime specifies source specific standards and relies on regulatory instruments for enforcement of the standards. For non-point sources like vehicles, measurement and monitoring of emissions for vehicles in use involve high transaction costs. Hence, we need indirect instruments to prevent and control pollution.

In this chapter, we explore the feasibility of using a mix of economic and regulatory instruments to tackle vehicular pollution in India. Section 2 deals with the growth of automobiles in India, contribution of vehicular pollution to air pollution, initiatives taken by Government to prevent and control vehicular pollution, and the case for introducing ecotaxes on automobiles. Section 3 considers the basis for setting the ecotaxes in India and explains the three ecotax proposals given in the Interim Report. Section 4 summarises SIAM's comments on the three ecotaxes and our responses and modifications to the proposals. We also examine the issue whether pollution from diesel is more harmful than that of petrol. Section 5 suggests policy instruments for pollution control from vehicles-in-use. Section 6 contains the recommendations.

Vehicular Pollution in India

Growth of Automobiles

There has been a rapid increase in the number of registered motor vehicles in India, from 0.306 million in 1951 to 58.863 million in 2002. See Table 4.1. The number of vehicles per 1000 population increased from 0.83 in March 31, 1951 to 2.53 in March 31, 1991 and to 5.35 in March 31, 2002. The compound annual rate of growth during 1990-91–2001-02 was 9.2 per cent.

There has also been a significant change in the composition of registered vehicles. Two-wheelers, cars, jeeps and taxies accounted for 60.8 per cent of the vehicles in 1951 but the percentage increased to 83.3 in 2002. During 1990-91–2001-02, the compound annual rate of growth was the highest for two wheelers (9.7 per cent), followed by cars, jeeps and taxis (8.6 per cent), others including tractors, trailors, three-wheelers and other miscellaneous vehicles (8.0 per cent), goods vehicles (7.4 per cent), and buses (6.4 per cent). In the absence of public policies to encourage mass passenger transportation, the shares of two wheelers and cars will increase over time with the higher rate of growth in GDP and also preferences toward personal mode of transportation.

Table 4.1

Total Number of Registered Motor Vehicles in India—1951-2002

(in thousands)

Year (As on 31 March)	*All Vehicles*	*Two-wheelers*	*Cars, Jeeps and Taxies*	*Buses*	*Goods Vehicles*	*Others**
1	2	3	4	5	6	7
1951	306	27	159	34	82	4
1956	426	41	203	47	119	16
1961	665	88	310	57	168	42
1966	1099	226	456	73	259	85
1971	1865	576	682	94	343	170
1976	2700	1057	779	115	351	398
1981	5391	2618	1160	162	554	897
1986	10577	6245	1780	227	863	1462
1991	21374	14200	2954	331	1356	2533
1996	33786	23252	4204	449	2031	3850
1997	37332	25729	4672	484	2343	4104
1998	41368	28642	5138	538@	2536	4514
1999	44875	31328	5556	540@	2554	4897
2000 (R)	48857	34118	6143	562@	2715	5319
2001 (P)	54991	38556	7058	634@	2948	5795
2002 (P)	58863	41478	7571	669@	3045	6100

Note: * : Others include tractors, trailors, three-wheelers (passenger vehicles) and other miscellaneous vehicles which are not separately classified.

@ : Includes omni buses.

(P) : Provisional

(R) : Revised.

Source: Ministry of Road Transport and Highways, *http://morth.nic.in/mts.htm.*

Share of Vehicular Pollution in Total Air Pollution

In the absence of adequate data on emission inventory, the Expert Committee on Auto Fuel Policy, Government of India (2002), could not quantify the contribution of automobile exhaust to the different pollutants in the air for the country. After reviewing some air quality studies undertaken by different agencies from 1988 to 1995, it summarised the contributions of transport, industry and other sectors to air pollution in Delhi and Mumbai. As is shown in Table 4.2, the transport sector is the major contributor of CO and NO_x both in Delhi and Mumbai. As for SO_2 and PM, the contribution of the transport sector is relatively small. The Report also finds that in almost all the eight major cities, the traffic load in terms of vehicle kilometres per day is mainly (70-80 per cent) constituted by private automobiles (cars and two-wheelers). In each of the eight cities the largest single contributor to the traffic load is two-wheelers (p.72).

The Report notes that the contribution of auto exhaust to overall pollution depends on factors such as the state of in-use vehicles and their relative population, availability, and effectiveness of the inspection and maintenance system, traffic management, status of roads and availability of bypasses for long distance traffic, land use pattern including siting of the parking depots, fuel quality and the pricing system for fuels (p.27).

The share of new vehicles introduced every year is about one-tenth of the stock of vehicles in use. In general, emission per vehicle increases with its age. The Society of Indian Automobile Manufacturers (SIAM) (2003) reports that significant emission reduction is possible if old vehicles are retired. From a study of vehicular emissions in the six largest cities in India, it finds that the percentage emission reduction if pre-1986 private vehicles are retired ranges from 25.3 in Ahmedabad to 73.8 in Mumbai; the corresponding percentage for retirement of pre-1986 commercial vehicles ranges from 30.89 in Mumbai to 64.61 in Kolkata. Hence, for control of vehicular policy we need policies for improvement in new vehicle quality, improvement in fuel quality, monitoring auto exhausts of in-use vehicles and retirement of very old vehicles.

Table 4.2

Ambient Air Quality in Delhi and Mumbai Pollution by Sources

(per cent)

Pollutants	*Sectors*		
	Transport	*Industry*	*Domestic & Others*
		Delhi	
CO	76-90	37-13	10-16.3
NO_x	66-74	13-29	1-2
SO_2	5-12	84-95	0-4
PM	3-22	74-16	2-4
		Mumbai	
CO	92	8	0
NO_x	60	40	0
SO_2	2-4	82-98	0-16
PM	0-16	34-96	53-56

Note: This summary is based on air quality studies undertaken by different agencies from 1988 to 1995 as given in Table 4.6 of Government of India (2002), *Report of the Expert Committee on Auto Fuel Policy*, pp.44-45.

Initiatives Taken to Control Vehicular Pollution

The Government of India has taken a number of steps to control vehicular pollution. The initiatives are:

- Introduction of compressed natural gas (CNG), as cleaner auto fuel in Mumbai, Delhi, Baroda and Surat.
- Improvement in petrol quality.
  - Phasing out of lead from April 1, 2000.
  - Increase in octane number with effect from April 1, 2000.
  - Reduction in sulphur content from 0.20 per cent maximum to 0.10 per cent maximum from April 1, 2000; in the four metros sulphur content to a maximum of 0.05 from April 1, 2005.
  - Benzene content limit from April 1, 2000: 3 per cent volume maximum for four metros and 5 per cent volume maximum for the rest of the country.
- Improvement in diesel quality.
  - The sulphur content in diesel has been reduced from 1.0 per cent maximum in April 1, 1996 to 0.25 per cent

maximum in April 1, 2000. In the four metros sulphur content has been reduced to 0.05 per cent.

* Increase in octane number from 45 to 48 from April 1, 2000.
* Improvements in distillation specification of diesel.

The Expert Committee on Auto Fuel Policy has laid down a road map for vehicular emission norms for new vehicles. These norms are as follows:

A. New Vehicles (except two and three wheelers)

Entire Country

* Bharat stage II emission norms from April 1, 2005.
* Euro III equivalent emission norms from April 1, 2010.

For cities of Delhi/NCR, Mumbai, Kolkata, Chennai, Bangalore, Hyderabad, Ahmedabad, Pune, Surat, Kanpur and Agra.

* Bharat stage II emission norms.
* Delhi, Mumbai, Kolkata and Chennai (in 2000 and 2001).
* the other cities from April 1, 2003.
* Euro III equivalent emission norms for all private vehicles, city public service vehicles and city commercial vehicles from April 1, 2005.
* Euro IV equivalent emission norms for all private vehicles, city public service vehicles and city commercial vehicles from April 1, 2010.

B. New Two and Three Wheelers in the Entire Country

* Bharat stage II norms from April 1, 2005.
* Bharat stage III norms preferably from April 1, 2008 but not later than April 1, 2010 in any case.

The details of vehicular emission norms are given in Tables 12.3 to 12.6 of the Report (pp.187-190).

As a result of these initiatives, the contributions of lead and SO_2 to auto exhausts will decrease considerably; the per vehicle contributions of CO, NO_x and SPM will decrease but their aggregate contributions may decrease only slowly because of the high rates of growth of vehicular population.

Case for Introducing Ecotaxes on Automobiles

The Expert Committee on Auto Fuel Policy has given a road map for attaining vehicular emission norms and made many suggestions for reduction of pollution from in-use vehicles (Chapter 14). The Committee suggests that, "the Auto Fuel Policy should be designed to set in motion a process of introducing in the country, economic instruments that would lead to improved air quality" (p.228). The Committee notes the distortion in fuel pricing, particularly higher relative price of petrol compared with diesel and "non-inclusion of high emissions/external cost of vehicles converted to diesel" (p.230). The passenger cars segment of the Indian automobile industry bears a higher incidence of excise tax compared with the other segments, the excise duty on them being 24 per cent while the rate for the other segments is only 16 per cent. All segments have to pay local level taxes and duties. The present tax policy does not take into consideration either energy efficiency or the external costs of using petrol and diesel by vehicles.

Since vehicular pollution is a non-point source of pollution, there is scope for using taxes on inputs and outputs for improving environmental quality. Such taxation would include higher tax rate on fuel inefficient vehicles, higher tax rate on polluting vehicles and on fuels not meeting the desired environmental norms. It may also be noted that petroleum is an exhaustible resource and India, at present, meets 73 per cent of its petroleum demand *via* imports. Hence, some discouragement for the use of hydro-carbons is in order.

In many OECD countries, taxes on motor vehicles have been restructured to incorporate environmental concerns, while also fulfilling the basic objectives of revenue raising. For example, in Germany an annual tax is levied on motor vehicles not meeting the EU emission standards in order to accelerate the introduction of cleaner vehicles. Rates are differentiated by age of the car. In Denmark an annual tax on passenger cars is levied according to the fuel economy of vehicle. The tax rates vary between 440 and 16100 DKK per year for petrol driven vehicles and between 1720 and 22020 DKK per year for diesel driven passenger cars. The Netherlands has lower sales tax on cars that comply with future EU standards. USA has a gas guzzler tax.

Details regarding application of fiscal instruments for controlling vehicular pollution in various countries are given in Annexure A-4.1.

Ecotax Proposals in the Interim Report

Ecotaxation–Basis

In theory, the tax should be equal to the marginal environmental damage costs. An adequate estimate of marginal environmental costs is difficult to obtain. There exists only one example (the UK landfill tax, introduced in October 1996), which is explicitly based on an estimate of the damage costs. In view of the non-availability of reliable estimates of marginal environmental damage costs, incentive effects and revenue raising become underlying criteria in setting environmental taxes. According to Ribeiro *et al.* (1999), two other criteria may be substituted for the theoretical criterion of equality of tax rate and marginal environmental damage costs. These are environmental effect (i.e., the effect of tax on environmental pollution) and incentives effect (i.e. a comparison with the marginal pollution-abatement costs, or as a proxy, the average abatement costs of the measures taken by the taxpayers). The first criterion directly attempts to trace the contribution of the tax to the monitored pollution reduction, while the second criterion endeavours to establish incentives for taxpayers to change their behaviour in a way more favourable to the environment often by adopting abatement measures.

Since the road map of emission norms has been set, the issues in emission performance of new vehicles would be taken care of. It may, however, be noted that two vehicles of the same class/type (say passenger cars) may confirm to a given emission standard (specified in grams per kilometre), say Euro II, yet total emissions of these vehicles may be at variance if fuel consumption/km of these vehicles varies. The improvement in fuel economy of vehicles results in burning of less fossil fuel (fuel conservation). This in turn, has a bearing on green house gas emissions as well as on regulated emissions. Thus, the fuel economy of vehicles is of great significance and most of the vehicle manufacturers world over are undertaking research and development to achieve better fuel economy which also helps in developing a competitive edge in the market place.

In USA, the EPA and the Department of Energy, each year publishes a fuel economy guide listing the fuel economy of new passenger vehicles. Manufactures are required to place a window sticker-containing city and highway fuel economy values of all new cars and light trucks (less than 8500 pounds of GVWR) when they are offered for sale or lease. At present, although no member of EU applies fuel economy standards the European vehicle manufacturers' association (ACEA) has made a commitment to reduce the fuel consumption of new vehicles voluntarily, to support the EU's Kyoto commitment to avoid application of regulatory or fiscal policies to vehicle fuel consumption.

Setting the Rate of Ecotax

Thus, we see that there is considerable global evidence on the use and application of economic instruments in addressing the issues of emission and fuel consumption of vehicles. In view of the wide use of taxes and subsidies in providing incentives to use fuel efficient vehicles there is a strong case to consider the feasibility of introducing incentives for improving fuel economy in the Indian context. The incentive structure can be linked to (a) the cost of upgrading the engine technology and (b) the benefits derived from such upgradation, which is the health damage cost foregone due to less fuel burnt per km and resultant lower emissions. The incentive structure can also be a weighted average of both the above factors. However, there is very sketchy or no information available on both aspects. As far as vehicle technology is concerned, there is no information available in the public domain on cost requirement to bring about better fuel economy for different models and category of vehicles.

The calculation of benefit in terms of health cost foregone is a problem due to non-estimation of dose-response functions in the Indian vehicular pollution context through epidemiological studies. Indian data in regard to the proportion of vehicular emissions related morbidity and its impact on health are insufficient. Therefore, a differential tax structure is being suggested based on the fuel economy of different types of vehicles complying with various Euro norms.

Only private vehicles have been considered for ecotaxation in this chapter. This is, mainly because commercial vehicles such as buses, trucks, three-wheelers and light commercial vehicles have only a few models and these fall within a narrow band of variation in terms of fuel economy. Further, in the case of heavy-duty vehicles, fuel economy data in km/litre will not be available as the heavy engines are tested separately on engine dynamometer and fuel consumption is recorded in g/kwh. In private vehicle category, two-wheelers, cars and jeeps have been considered. Although fuel economy achieved by Indian two-wheelers is comparable/superior to that achieved in other countries for the same class of vehicles, there is a lot of scope for improvements in view of the large population of this class of vehicles in India.

Data on fuel economy of vehicles have been obtained from SIAM. Fuel economy data are based on type approval test conducted under controlled environment. On road, fuel consumption of vehicles would depend upon factors such as traffic conditions, vehicle condition, driving habits. Since the proposed ecotax directly targets the fuel consumption of vehicles it may be more appropriate to call it a resource tax. In addition to conservation of resource, the resource tax would have an impact on emissions of both local and global pollutants. Resource tax factor has been derived on the basis of categorisation of vehicles based on fuel economy. SIAM has provided data for a few makes of two and four-wheelers. These data are for Bharat stage II vehicles. Data on passenger cars and jeeps show strong negative correlation between engine capacity and fuel economy of the vehicles, indicating that the fuel economy of vehicles decreases with an increase in engine capacity of the vehicles. The following main observations can be made on the basis of data provided by SIAM:

(i) Vehicles reporting fuel economy <10 have engine capacity $\geq$ 2000 cc.

(ii) Of the four makes of vehicles falling in fuel economy range of 10-12, three have engine capacity $\geq$2300 cc. One has engine capacity of around 1800 cc. This suggests that the vehicle technology is an important determination of fuel economy although engine capacity of vehicles also has a strong bearing on their fuel economy.

(iii) A large number of vehicles are in the fuel economy range of 12-14 km/l. Many of these have engine capacity $<$ 1800 cc. Some vehicles, however, have higher engine capacity.

(iv) Fuel economy category of 14-16 km/l also has a large number of vehicles. Most of these however have engine capacities $<$ 1500 cc.

There is a large concentration of vehicles in the fuel economy range of 12-16. Engine capacity of most of these cars range between 1000 cc to 1800 cc. The trend towards higher engine capacity is already evident in the figures for FY 2003. This trend is likely to accelerate in the medium term. Eventually, the 1000-1700 cc segment is likely to dominate the Indian car market (the combined share of the total sales in this segment increased from 44.5 per cent in FY 2001 to 63.3 per cent in FY 2003), (ICRA, 2003). Since the regulation for enhanced environmental compliance is already in place, environmental performance is likely to be the major factor prompting change in automobile technology in India. This is perhaps the right time to set the target for improvement in fuel economy of vehicles by sending appropriate signals through a resource tax on vehicles.

Rates of Resource Tax

As the proposed ecotax directly targets the fuel consumption of vehicles it may be more appropriate to call it a resource tax. In addition to conservation of the exhaustible resource, the resource tax will reduce the emissions of local pollutants and global pollutant (CO_2). The proposed resource tax rates for two-wheelers, passenger cars and jeeps are given in Table 4.3. Based on fuel economy slabs the resource tax factors are arrived at. For two-wheelers, the slab 61 km/l and above has a resource tax factor of 1 and for the slab up to 40 km/l the resource tax is 1.75. There is no resource tax for the most fuel-efficient vehicle. The vehicle resource tax rate increases as the fuel economy decreases.

For passenger cars and jeeps, the slab more than 20 km/l has a resource tax factor of 1 and for the slab up to 7 km/l the resource tax factor is fixed at 2.85. The best fuel economy passenger cars and jeeps would be exempt from the resource tax. For the least fuel-efficient category the tax rate is 10 per cent. The bases of resource tax would be the prices of various categories of vehicles before CENVAT.

Table 4.3

Resource Tax Rates

Vehicle	*Fuel Economy Km/1*	*Resource Tax Factor*	*Rate of Resource Tax (per cent of vehicle price before excise tax)*
Two-wheelers	Up to 40	1.75	5.0
	41-50	1.4	3.0
	51-60	1.16	1.5
	Above 60	1.0	0
Passenger Cars and Jeeps	Up to 7	2.85	10
	8-12	1.66	5
	13-16	1.25	3
	16-18	1.11	2
	Above 20	1.0	0

Petrol versus Diesel Cars

Statistics on passenger vehicles show that in recent years new models of diesel cars have been introduced in the market and that there is a significant rise in the growth of sale of diesel cars. This trend is attributed mainly to the significant difference in petrol and diesel prices—a result of the governments' fuel pricing policy that has evoked considerable discussion on the desirability of diesel versions of passenger cars in India. The main concern is the reported greater pollution potential of diesel vehicles *vis-à-vis* petrol vehicles.

Emission rates of sulphur oxide (SO_x), nitrogen oxides (NO_x) and particulate matter (PM) are higher in diesel vehicles *vis-à-vis* petrol vehicles fitted with catalytic converters. Relatively lower emissions of CO and HC *vis-à-vis* petrol vehicles, fuel efficiency, longer engine life are some of the advantages of diesel vehicles. These are, however, not the main reasons for the spurt in supply and demand for diesel vehicles in India. The key factor responsible for this is the price difference between diesel and petrol. In other countries, where diesel does not enjoy a significant price advantage *vis-à-vis* petrol, growth in demand for diesel vehicles is mainly driven by two factors; (i) given the vehicular engine technology, diesel engines are less costly to adapt than petrol engines, and (ii) diesel engines are more fuel efficient. Experts say that the problem of PM emissions has to a large extent been solved with the development of a device called CRT for diesel vehicles and there is a lot of scope for improvement in this technology to further reduce the harmful emissions of diesel vehicles.

It may not be prudent to ban non-commercial diesel vehicles. Instead, the government should end the price discrimination in favour of diesel and against petrol to discourage excessive use of diesel vehicles because of the artificial fuel price advantage. The complete deregulation of prices of hydrocarbons will resolve part of the price distortion. If the artificial price difference between petrol and diesel is removed only in a phased manner, then to avoid the environmental damage due to diesel vehicles in the interim, the regulator should levy a tax on diesel vehicles that would induce motorists to consider alternative fuels. Such a tax should be based on the average cost that emissions from diesel vehicles inflict on the society. It may be noted, that petrol driven vehicles also emit pollutants in the atmosphere that impose a cost on the society. Since emissions from petrol vehicles are not currently taxed in India, the base for a tax on diesel vehicles should be the difference between average costs of emissions from diesel vehicles and petrol vehicles. Information on the cost of emissions is however not available for all the pollutants. In the context of India, some estimates of health costs of PM_{10} are available (Brandon and Homman, 1995). We thus take the emission of PM_{10} as the additional negative externality from diesel vehicle *vis-à-vis* a petrol driven vehicle.

Following the Brandon and Homman (1995), the present value of cost of pollution (income loss and medical costs) due to emissions by a diesel car over a period of 10 years[1] is estimated to be Rs. 10,648. This works out to Rs. 1,109 per year. An annual emission tax of Rs. 750 (an average of lower and upper bound estimates of annual costs due to PM_{10}) should be levied on diesel cars. The proposed tax rate may, however, be lower than the tax rate that would be obtained at the intersection of the marginal abatement cost and marginal damage cost curves.

It must be noted that this measure may not have the desired impact as long as difference between diesel and petrol prices is significant. But it may be expected that the distortion in the relative prices of petrol and diesel will be phased in course of time. We shall not therefore try to correct its impact. Dealing only with the environmental impact difference, we recommend that an annual emission tax of Rs.750 may be levied on each diesel car.

Comments on the Interim Report by SIAM and Modifications in the Proposals

SIAM's Comments

SIAM presentation at the Ministry and in its letter addressed to Dr. Chelliah dated[1] May 1, 2003 had objections to all the three proposed ecotaxes. Regarding the resource tax it objects to the tax because (a) the classifications of vehicles into different km/litre slabs is arbitrary, (b) fuel consumption level varies from variant to variant depending on the consumer profiles, and (c) wide variation in fuel efficiency of the same model tested at different times. Regarding the emission tax for diesel cars, its view is that with past APM dismantling, the disparity in price is reducing. Regarding the tax on

1. Life of cars is estimated to be 10 years.

diesel car it questions the assumption that diesel is more polluting than petrol. It states that European markets are steadily shifting towards diesel cars because of the inherent fuel efficiency and that a diesel tax will retard the development of the technology in India.

SIAM is against any new fiscal measures because the tax burden on passenger cars is high (34.6 per cent of consumer price), passenger car penetration is low in India (6 per 1000 in 2001) and new taxes will affect not only the industry but the economy as a whole because of the industry's backward and forward linkages to other sectors. It feels that "the best option would be for competitive environment determining market forces where customers should have informed choice for decision making."

Modifications in the Proposals

Our response is that, in the presence of negative externality (vehicular pollution), the unregulated competitive market outcome will not produce socially desirable outcomes. The tax burden in the auto industry, despite the reduction in excise duty from 32 per cent to 24 per cent on passenger cars and multiutility vehicles, is high compared with many other countries. We argue only for differential taxes with lower tax rates for eco-friendly items and higher tax rates for polluting items. In fact we have argued for tax neutral environmental reform in Chapter 2 and also supported the use of environmental cesses/charges to finance activities which improve environmental quality. We give the following revised proposals:

Resource Tax on Passenger Cars, Jeeps and Two-wheelers

There is a case for resource tax to promote energy efficiency. However, we simplify the tax structure and moderate the tax increase. Our objective is not revenue generation but energy conservation directly and reduction of air pollution indirectly.

Resource tax factor has been computed as follows. The resource tax factor for the best fuel economy category, is taken as 1 for both cars and two-wheelers. For the next fuel economy category, the resource tax factor for two-wheelers is calculated as 60/1x55 (where 55 is the average of the fuel economy range). Similarly, the resource tax factor for the next category of two-wheelers would be 60/45.

Best fuel economy category has been exempt from resource tax. For the worst fuel economy categories for two-wheelers and passenger cars/jeeps, the resource tax is proposed to be at 3 per cent and 5 per cent, respectively, on factory price of vehicles before excise tax (central) is levied.

The resource tax for the next fuel economy category for two-wheelers is calculated as (3x1.33)/1.5. The rate of resource tax in this case works out to 2.66. For ease of implementation this has been rounded off to 2.5 (column 4, row 2, Table 4.4). The rate of resource tax for other categories has also been rounded off to keep the tax rate structure simple.

The rates indicated below clearly show what is aimed at is differential taxation to induce the incidence effect as well as higher rate corresponding to higher cost. Since the CENVAT rate is already high, the ecotax resource tax rates are pitched low.

Table 4.4

Rates of Resource Tax

Vehicle	*Fuel Economy*	*Resource Tax Factor*	*Rate of Resource Tax (per cent)*
Two-Wheelers	0-40	1.5	3.0
	41-50	1.33	2.5
	51-60	1.09	2.0
	61 and above	1	-
Passenger Cars/Jeeps	<10	1.67	5
	10-14	1.25	3
	12-18	1.11	2
	> 18	1	-

Annual Emission Tax on Diesel Cars

Our proposal of an annual emission tax of Rs.750 on diesel car is based on estimates of income loss and medical costs attributable to particulate matter. Ideally, the emission tax on diesel car should be based on the difference in particulate emission between a diesel car and a petrol car. At present, there is no standard for particulate emission in petrol car. Since particulate emission from a petrol car is not zero, the proposed emission tax has an upward bias. We agree that there are difficulties in deciding the tax rate based on one pollutant (PM_{10}) and result based on the US data. Many including

Copper, Simon, Alberni, Sharma[2] argue that the extrapolation of results from studies conducted in industrialised countries about developing countries are inappropriate because (a) the relationships found between pollution and health at the low levels of pollution experienced in industrialised countries may not hold for the extremely high pollution levels witnessed in developing countries, and (b) extrapolation of the impacts of air pollution on mortality may be especially misleading because people die at younger ages and from different causes in developing countries than in industrialised countries.

The second objection is that as both petrol and diesel cause pollution and as the pollution vectors for petrol and diesel are different, one must arrive at an appropriate weighting scheme to estimate the overall pollution load from petrol and diesel. Sengupta *et al.* (2002) undertook such a study. They made corrections to take into account differences in temporal and cross country situations between USA and India. Based on their estimates, Mukherjee and Sud (2003) calculated the health costs imposed by a passenger car, when it runs for 1000 kilometres in Indian cities, for diesel and petrol cars for different Euro norms.

Table 4.5

Health Costs in Rs./1000 Km in 2000-01 Prices for Diesel and Petrol Cars

Standard Pollutants	*EURO I*		*EURO II*		*EURO III*		*EURO IV*	
	Diesel	*Petrol*	*Diesel*	*Petrol*	*Diesel*	*Petrol*	*Diesel*	*Petrol*
CO	0.27	0.27	0.10	0.22	0.06	0.23	0.05	0.10
HC	—	—	—	—	0.09	0.29	0.07	0.14
NOx	—	—	—	—	11.25	3.38	5.63	1.80
HC+NOx	23.22	23.22	16.76	11.97	—	—	—	—
PM	25.44	0.00	14.53	0.00	9.08	0.00	4.54	0.00
Total	**48.92**	**23.49**	**31.39**	**12.19**	**20.49**	**8.30**	**10.29**	**4.25**

Source: Mukherjee and Sud (2003), Table 4b, p.41.

2. See, *http://www.worldbank.org/html/dec/publications/workpapers/WPS/1800series/wps1860.pdf.*

The above results show that diesel cars are more polluting than petrol cars, but the differences in health costs are decreasing as one moves from Euro I to Euro IV.

There is controversy whether diesel is more polluting than petrol. Advocates of diesel cars argue that diesel cars are more fuel-efficient than petrol cars and with rapid advances in diesel technology the health costs (due to PM) will decrease. There is also a changing perception about particulate matter. It has been found that it is mainly fine particles ($PM_{2.5}$: 2.5 mm) rather than PM_{10} which is the main cause of exposure related health hazards. It is contended by some that the number and size are more important than the mass for assessing the health effects. See Mukherjee and Sud (2003), a study commissioned by MSE, for a comparative assessment of the impact of emissions from diesel and petrol fuelled passenger cars. That study refers to the evidence on fine particles emitted by petrol vehicles. *Down to Earth* magazine which takes the position that diesel cars are more polluting than petrol cars, in a cover story of its issue dated September 15, 2003 reports that nanoparticles, as tiny as a billionth of a metre, pose a challenge to scientists and regulatory authorities alike. It says that research on nanoparticle emissions from vehicles is still in the academic domain (p.31). While our recommendation for a discriminatory tax on diesel vehicles is justified taking into all factors, the government may carry out comprehensive examination of the latest findings of the emissions from petrol and diesel vehicles before imposing the tax.

In-use Vehicles

It is being increasingly realised that despite rapid improvement in emission standards for 'new vehicles', emissions from automobiles will continue to be a major source of urban air pollution in the large cities in India. The reasons for this are complex and numerous, but two important factors are the deterioration in performance of emissions control equipment and other systems on the vehicle that affect emissions as such vehicle's age, and the number of older vehicles with less effective pollution controls which are still in use.

Policy Instruments for Pollution Control from Vehicles in Use

In addressing the pollution problems from in-use vehicles the following policy instruments have been discussed in the literature.

Regulatory Programmes

Regulatory programmes involve compulsory periodic inspection and testing of motor vehicles and require owners to repair failing vehicles, to enforce compliance with applicable emission standards. These programmes put the onus of bringing the vehicle in for testing, as well as the cost of any repairs that might be necessary, on the motorists. Fixing the responsibility of testing and repairing on owners seems to make sense for at least two reasons. First, some repairs that reduce emissions might also result in better driveability and better fuel economy that vehicle owners should care about. Second, making motorists responsible is consistent with the 'polluters pay' principle. Periodic Inspection and Maintenance programme (I/M) has been implemented in a number of countries (Annexure A-4.1). In the USA, I/M programme is not only in operation for three decades but has also been examined to understand the costs and effectiveness of this programme. For a review of studies, which examined I/M performance, see Harrington *et al.* (1999).

I/M programmes were first introduced in the US in the late 1970. It has been reported that these programmes were not very effective as actual pollution reductions were very small and total costs of I/M programme were very high.[3] According to Harrington *et al.* (1999) I/M has been a disappointment. The study states that, "our review of the existing in-use emission reduction programmes persuades us that I/M will continue to disappoint as long as motorists are held responsible for the emissions of their own vehicles." The paper claims that this conclusion is shared to a considerable degree by the regulators, who have responded by forcing the manufacturers to extend the warranty provisions on new vehicles; in effect pressuring manufacturers to reduce emission rates in new vehicles and make emission systems in new vehicles more durable and impervious to poor maintenance by the owners. The paper further states that this approach is not likely to be efficient, at least in the short run, because it does not do anything about the emissions of existing

3. Ando *et al.* (1998) which examines I/M programme in America shows that only 29 to 38 per cent of the total costs of the programme is devoted to the repair of the vehicles while the rest is used for vehicle emission testing. Transaction costs also arise because motorists have ample opportunities for evading the responsibilities. Motorists can fail to take the emission tests, they may opt for incomplete report: they may register their vehicles outside the I/M jurisdiction while continuing to use it there, or they may fail to register their vehicles at all.

vehicles and thus it will take a very long time to have an appreciable effect on fleet emission. Harrington *et al.* (1999) explores alternative liability assignments for in-use vehicle emissions and examines the kind of policies that would be necessary to change those assignments and compares how they affect the transaction costs associated with existing I/M programmes (Table 4.6).

As shown in Table 4.6, each of the discussed instruments address some of the transaction cost categories characteristic of I/M. In particular, by shifting responsibility from motorists, all instruments remove the motorists' incentive to avoid emission repair.

Emissions Fee

Emission fee on vehicles has been suggested by many economists (White, 1982; Harrington *et al.*, 1996; Eskeland and Devarajan, 1996; Smith, 1995). A pollution fee programme involves charging a fee from vehicles based on their emission rate (grams per km) times total kms driven. According to Harrington *et al.* (1996), such an approach (with an I/M programme used only for determining the emissions rate) has the potential to substantially increase economic welfare over regulatory I/M policy (as currently in force in the US) for reducing vehicle emissions for at least two reasons. First, under the regulatory I/M policy motorists must repair a vehicle if it does not pass the inspection, regardless of the benefits of doing so whereas with an emissions fee motorists have a choice about whether to repair the vehicle or to avoid costly repairs if they are unlikely to produce significant emission reductions. Second, an emissions fee can especially target vehicles that contribute the most emissions, i.e., that have a high emissions rate and high mileage, whereas I/M programmes treat all dirty vehicles equally.

Besides the difficulties in determining the rate of emissions fee and in its implementation, the fee system is criticised for being unfair to low income earners (because they are likely to disproportionately own dirty vehicles). To partly address this and to raise the general political viability of an emissions fee system, it has been suggested that the revenues be recycled to citizens of the affected area in such a way as not to interfere with the incentives to reduce polluting behaviour. Harrington *et al.* (1996) using a simulation model based on empirical evidences about vehicle emissions and repair effectiveness compares I/M programmes in the US with various

Table 4.6

Alternative Approaches to Sharing Emission Liability: Summary of Characteristics

Policy	*Transaction Cost Category*			*Manufacturer Incentives to Produce Durable Vehicles*
	Motorist Avoidance Incentives	*Monitoring Costs*	*Transferability of Emission Reductions*	
Extended warranties[a]	Depends on length of warranty. Some effort required to ensure motorists bring in vehicles to be repaired.	No effect on existing vehicles; new vehicles could be identified with OBD.	Some within-manufacturer transferability possible if emission averaging is allowed. But potential savings are low.	Strong
Motorist subsidies or emission repair insurance[b]	Depends on subsidy levels. If high and broad, there will be little incentive to avoid repair.	Substantial cost reduction if gross emitters can be identified by remote sensing or OBD.	Difficult to overcome agency problems if private mechanic are given discretion over repair decisions.	Weak
Centralised repair liability[c]	Minimal incentive to avoid repair.	Public costs could be low, since average sub fleet-specific emissions can be estimated by RSD. Private costs are uncertain and depend on the case of winning motorists' cooperation.	Emission reductions are quite transferable. Well-adopted to emission fee or tradable permit regimes.	Weak, unless manufacturers are parties responsible for repair.
Mandatory leasing[d]	Minimal incentive to avoid repair	Lowest costs.	Emission reductions are quite transferable. Well-adopted to emission fee or tradable permit regimes.	Strong

Note: The baseline for the implied comparisons in the table is the Enhanced I/M programme as promulgated by EPA in 1992. Other Comments:

a: Only applies to new vehicles; therefore no immediate effect on emissions. b: Public funding source required. c: Political problems if liability for groups of vehicles is assigned by fiat. High administrative start up costs if liability is auctioned off. d: Possibly serious political opposition. OBD: On-board devices that continuously check emissions systems and signal motorists if controls are not working properly. RSD: Remote-sensing devices.

Source: Harrington *et al.*, (1999).

emissions fee policies. The difference between economic incentive policies and regulatory I/M policies would depend on how well the incentive polices are targeted and on their ability to influence polluter behaviour. The vehicle emissions tax considered in Harrington *et al.* (1996) is a tax rate on grams of pollutants per mile. A vehicle owner can respond to an emissions fee by repairing the vehicle, by driving it less, or by scrapping it.[4] The paper finds that fees have higher net benefits than the regulatory I/M policy. It also finds that the extent of uncertainty in measuring and repairing vehicle emissions can have important impact on emission reduction potential of both regulatory as well as emissions fee policy. The regulatory policy is particularly affected by uncertainty, as there is limit to the amount of possible emissions reductions. Under uncertainty, a fee which only has to be paid after emissions reach some limit, is found to be as efficient as pure fees[5] and has the advantage of mitigating the high costs of fees to motorists.

This chapter is mainly concerned with ecotaxes that could be levied by the central government and also deals only with its areas of responsibility. The large area of responsibility of dealing with the in-use vehicles and regulating their use falls in the jurisdiction of state governments. It is well known that the environmental damage caused by automobiles increases with their age and length of use. The damage is seen to increase progressively with age. Hence those who use older vehicles have to pay for greater damage and it is also desirable to encourage them to scrap the old vehicles. Several measures to achieve these objectives have been discussed in the literature[6] and also tried in a number of countries. These kinds of measures fall outside the scope of this chapter. However, a separate chapter can be written on ecotaxes that could be levied and other related measures that could be implemented by the state governments to deal with the environmental effects of in-use ageing vehicles.

---

4. The paper focuses on vehicle repair.
5. Two types of fee are considered. The first type has no exempt emissions, i.e., a fee to be paid on all emissions or baseline = 0. In the second type, baseline= 1, owners must pay fee only on emissions exceeding the baseline. Pure fee refers to the former.
6. See for example Alberini *et al.* (1998); Bellomo and Liff (1984); Cropper *et al.* (1997); Pandey (1988, 2004); Xie *et al.* (1998).

Recommendations

We recommend resource tax on passenger cars and two-wheelers at the rates suggested in Table 4.4 to encourage fuel economy. We also recommend an emission tax on diesel cars, however the government could carry out a resurvey of the evidence on the higher polluting effect of diesel cars before taking action.

Annexure A-4.1

Application of Fiscal Instruments in Various Countries

	Vehicle	
	Direct	*Indirect*
Australia		• Tax on sale or initial registration of vehicles. Rate of tax varies on the basis of value of vehicles. • Higher annual registration fee on commercial vehicles compared to private vehicles.
Austria		• Environmental tax on car registration was introduced in 1992. While the base is the selling price of cars the tax rate depends on the standard petrol consumption. • Since May 1, 1993, the annual vehicle tax on passenger cars is assessed on the basis of engine power and no longer on cylinder volume. From January 1, 1995, cars without catalytic converters are imposed a surtax of 20 per cent. • At the same time VAT on new vehicles was reduced from 32 per cent to 20 per cent and the VAT rate on electric cars was cut by half to 10 per cent. • Tax is based on net weight for buses and on loading capacity for trucks.
Belgium		• An annual tax on registration was introduced for new motor cars on June 1, 1992, and it has been extended to in-use cars since June 1, 1993. This tax is based on the engine power of the car. • Road tax on power(passenger transport vehicles) or weight(goods transport vehicles) adjusted annually on 1 July with CPI.
Brazil		• Penalty system for violation of air pollution standards since 1981. Fines are arbitrary as the level of emission from trucks is generally visually assessed. Fines are related to frequency of violation rather than intensity or toxicity of pollution.

contd. ...

...contd. ...

	Vehicle	
	Direct	*Indirect*
Britain		• Sales tax on new cars (17.5 per cent) and annual vehicle excise duty. • Higher taxes on commercial vehicle sales, ownership (excise duty based on axles and weight) and use than on private cars.
Canada	British Columbia introduced permit fees on pollutant emissions in 1992. The fees are reduced if the actual emissions are less than the permitted emissions.	• A tax on purchase of fuel inefficient passenger and sport utility vehicles. A subsidy of $100 is provided to cars with a highway fuel efficiency rating of less than 6 litres per 100 km. • All provinces impose annual fees for the registration or use of motor vehicles. In gen eral, the fees depend on the type of vehicles and in most cases the weight of vehicles.
Denmark		• Annual passenger car fuel consumption tax according to the fuel efficiency of vehicle. The tax rates vary between 440 and 16100 DKK per year for petrol driven vehicles and between 1720 and 22020 DKK per year for diesel driven passenger cars.
Finland		• Annual tax on diesel vehicles and passenger cars of 150 FIM/100 kg of weight, and on delivery vans of 27 FIM/1000 kg of weight. • Environmental taxes on cars, differentiated on the basis of whether or not these are equipped with catalytic converters. • Annual tax on diesel driven vehicles levied on all vehicles using fuel other than petrol. • Annual tax on motor vehicles levied on passenger cars and delivery vans 300 FIM/a for old vehicles(registr. before 1994) 500 FIM/a for new vehicles.
France		• Accelerated depreciations allowed for electrical vehicles.

contd. ...

...contd. ...

	Vehicle	
	Direct	*Indirect*
Germany		• Annual tax on motor vehicle not meeting the EU emission standards to accelerate the introduction of cleaner vehicles. Rates are differentiated by age of the car. • A higher tax has been imposed on diesel cars as compared to petrol cars since 1994.
Greece		• Since 1990 exemption from the road surtax and initial lumpsum tax for a period of 5 years for new cars fitted with a catalytic convertor, subject to scrapping of old car by the buyers of new car. About three lakh old cars were scrapped.
Hungary		• New cars with catalytic convertors get a discount of Forint 50000 from consumption tax.
Iceland		• IKr 330 is charged for a mandatory annual emission test. Inspection fee charged for annual inspection of vehicles over two years old. Rate of charge is on the basis of weight of vehicles.
Ireland		• Excise duty based on cylinder capacity of vehicles. • Sales tax on retail price of private vehicles based on cylinder capacity. • Road tax on private cars based on cylinder capacity.
Italy		• One-off registration tax on purchase of new and used vehicles depending on type and size of vehicles. • Annual tax on all motor vehicles and caravans. Levied in relation to cylinder vol. ume (light motorcycles), power (motorcycles, cars, coaches), load (lorries and trailers or lump sum (caravans). Power is measured in fiscal CV—an *ad hoc* unit based on cylinder volume. Further different rates of tax apply to petrol and gas oil motor vehicles. Tax rates are the sum of components for central and regional governments; the latter can be varied within limits by the regional authority and there is thus some regional variation in overall tax rates.

contd. ...

...contd. ...

	Vehicle	
	Direct	*Indirect*
Japan		• Tax deductions for cars with low emissions, electric cars and cars on alternative fuels. • One-off consumption tax on new or old car registration at three per cent. • Annual tax in relation to power and load of vehicles. • Automobile tax levied according to cylinder capacity. • Light vehicle tax levied on motor cycles and light vehicles according to cylinder capacity.
Korea		• Introduced environmental quality improvement charges, in 1991. Besides other economic activities it covered vehicles (buses and trucks using diesel). The charge is computed by the price of catalytic converter. • Vehicle excise duty levied according to type, weight, and fuel used.
The Netherlands		• Lower sales tax on cars that complied with future EU standards. Consequently, share of future EU standard cars rose from 37 per cent to 70 per cent, faster than expected by the government. • Vehicle excise duty levied according to type, weight, number of axles and fuel used.
Norway		• Differentiated taxes on car prices, with a tax advantage given to cars fitted with catalytic converters and cars powered by electricity or gas.
Portugal		• Vehicle excise duty levied on cylinder capacity.
Singapore		• Additional registration charge at 150 per cent, since 1983, of the cost of the car to discourage ownership. A rebate on tax is given if an old vehicle is scrapped simultaneously.

contd. ...

...contd. ...

	Vehicle	
	Direct	*Indirect*
Sweden		• Vehicle taxes based on weight and environmental characteristics(type, weight, number of axles and fuel used) • Subsidy on cars with catalytic converters, special tax on cars without catalytic converters.
Turkey		• Motor vehicle tax levied on all motor vehicles based on weight, type and cylinder capacity • Environment Fund at 25 per cent of the Supplementary Motor Vehicles Acquisition Tax.
United States		• Higher excise tax (12 per cent) on trucks, annual use tax on 'heavy vehicles', excise tax on tyres weighing over 40 pounds, a 'Gas Guzzler' tax on automobiles with unsatisfactory fuel economy ratings. • The Gas Guzzler excise tax is imposed on the sale of autos whose fuel efficiency is less than 22.5 miles per gallon. The tax varies from \$ 1000 to \$ 7700 depending on the fuel efficiency. • Non-conformance charge on heavy vehicles and engines are based on the degree of non-compliance.

contd. ...

...contd. ...

Country	Fuel	
	Direct	Indirect
Australia		• Petrol taxes were increased. Differential pricing for leaded and unleaded petrol was introduced in favour of the latter.
Belgium		• Higher excise duty on leaded petrol.
Britain		• Tax differential has been gradually increased and now stands at 4.8 pence per litre. The proportion of unleaded in total petrol sales rose to 50 per cent in 1993 from a negligible share in 1986.
Denmark		• Fuel tax based on CO_2 content at combustion was introduced in 1992.
		• Since the mid-1980s, differential tax on leaded and unleaded gasoline. In 1994 the market share of unleaded petrol rose to nearly 100 per cent.
Finland		• Lower tax on lead free petrol than on leaded petrol since 1986.
		• Lower excise duty on sulphur free diesel since 1993.
		• Carbon tax on fuel since 1994.
Germany		• Duty differential between leaded and unleaded petrol at the rate of DM 0.10 per litre.
Hungary		• In 1992, tax at the rate of 0.7 per cent of the price was introduced on motor vehicle fuels. The revenue is earmarked for environmental expenditure relating to vehicular traffic.
Ireland		• Higher excise duty on leaded petrol than on the unleaded petrol.
Luxembourg		• Higher excise duty and VAT rates on leaded petrol by 2-3 per cent than on unleaded petrol.
Mexico		• Higher excise tax on leaded petrol than on unleaded variety.

contd. ...

...contd. ...

Country	*Fuel*	
	Direct	*Indirect*
Netherlands		• Environmental charges on fuel since 1988. These charges were revised in 1990 to include CO_2 emissions. Unleaded petrol was cheaper than leaded petrol.
New Zealand		• Tax treatment in favour of unleaded petrol *vis-à-vis* leaded petrol. A fee of NZ$ 0.066 i.e. US$ 0.039 per gram was levied on lead added to gasoline.
Norway	Since 1995 gasoline tax difference was introduced for leaded petrol based on emissions of lead per litre.	• Fuel tax based on sulphur, carbon and lead content. • CO_2 tax since 1991.
Sweden		• High gasoline taxes. Differential tax in favour of unleaded petrol. • A carbon tax was imposed on motor and other fossil fuels since 1991. The part of the tax levied on motor fuels amounted to SKr 0.58 per litre for petrol and 0.92 for diesel. • System of tax rebate for producers of cleaner diesel fuel since 1991.
Switzerland		• The market share of unleaded petrol increased to 655 in 1992 due to a tax differentiation of ECU 0.04/l in favour of unleaded petrol.
Taiwan		• Differential price in favour of unleaded petrol. This led to an increase in the market share of unleaded petrol from 18.7 per cent in 1990 to 51.84 per cent in 1993.
Thailand		• Differential price in favour of unleaded petrol.
Thailand		• A surtax on leaded gasoline to finance the subsidy on unleaded gasoline.
US		• Trade in lead credits, to phase out lead in gasoline in 1982-87.

contd. ...

...contd. ...

Country	*Traffic*	
	Direct	*Indirect*
Chile		• In 1990 the city allocated bus transit rights and auctioned routes based on fares and types of buses.
Norway		• In 1986 to control congestion the city of Bergen introduced toll for motorists entering the city between 6 a.m. to 10 p.m. on weekdays. The rate is differentiated on the basis of loading capacity of vehicles. The revenue so collected is used to finance the construction of by-passes.
Singapore		• Implemented a licence ticket scheme for entering some identified zones during peak hours. Fine for non-compliance was 10 times the licence price. It helped reduce the traffic flow and thus pollution. These zones also have higher parking fee.

Source: Pandey and Bhardwaj (2000).

5

Chlorine Use in Pulp and Paper and Viscose Rayon Industries

Introduction

Chlorine is used as a bleaching agent in the pulp and paper industry because it is very effective in removing the brown colour in the pulp caused by lignin. However, chlorine particularly in its elemental form is very reactive and forms organochlorine compounds with the organic material in the pulp. The organochlorine compounds are discharged as part of the waste streams into the air or water and in some cases are embedded in the product. Organochlorine compounds resist natural breakdown processes and can build up over time in the environment. They enter the food chain through fish, meat, and dairy products. Organochlorines can also be found in the paper products themselves, and can leach into milk, coffee, and other foods with which they come in contact (Edwards and Santillo, 1996). Some of these compounds like dioxin are known to be toxic and can accumulate in the food chain.

There has been a global debate on the use of chlorine for bleaching and various technologies have been suggested for reduction in its use. Many countries have put restrictions on the use of chlorine and have also used 'ecolabelling' systems as a means of encouraging the use of chlorine free paper as well as recycled paper. Chlorine is also used in the bleaching of the pulp prior to the manufacture of viscose rayon. Viscose rayon plants, which use chlorine, discharge organochlorine compounds in the effluent which can ultimately enter the food chain. The concern about chlorine compounds is even greater in India because paper and pulp-mill effluents are often used to irrigate agricultural land. These persistent organic pollutants can accumulate in the soil or in the crops themselves.

Pulp and paper plants as well as viscose rayon plants use a lot of water and hence discharge a large volume of effluents. Typically, Indian pulp operations use as much as 35 cubic metres of water per tonne of paper. Since nearly 200 kg of chemicals are used per tonne of paper, the pollution load as expressed in terms of chemical oxygen demand (COD) and other parameters is quite high. The State Pollution Control Boards have the necessary powers and monitoring capability for enforcement of the standards for these parameters. Thus, the existing regulatory structures are adequate to deal with much of the water pollution caused by the pulp and paper industry. The major lacunae are the organochlorine compounds. Although the Central Pollution Control Board has promulgated a standard for organochlorine compounds (TOCL), the State Boards find it difficult to enforce the standard. An alternative to regulation is the use of economic instruments like ecotaxes to discourage the use of chlorine as a bleaching agent.

Industry Structure

The industry, based on raw material consumption, can broadly be classified as wood-based units, agro residue-based units and waste paper-based units. Similarly, on the basis of end-use, it can be classified as manufacturers of cultural paper, industrial paper, specialty paper and newsprint. While the country is self sufficient in cultural and industrial paper, it is largely dependent upon imports for newsprint.

The industry, with over 400 mills spread all over India, has an installed capacity of about 6.2 million tonnes. Of this, 36 mills are in the large-scale sector and the rest in the medium and small-scale sector. The industry is highly fragmented in that the installed capacities of the 36 large mills are almost equal to the installed capacity of 370 mills in small scale and medium units. However, 25 to 30 per cent of the industry's sales are from the top 10 producers.

Table 5.1

Key Statistics on Paper Industry

Indicator	*1995-96*	*1996-97*	*1997-98*	*1998-99*	*1999-00*	*2000-01*	*2001-02*	*2002-03*
Total Production (MMT)	3.55	3.91	4.29	4.45	5.09	4.81	4.95	5.24
Total Sales (Crore)	10,000	10,600	11,000	11,800	12,000	13,000	13,400	14,000
Export Quantity (MMT)	0.09	0.07	0.06	0.06	0.10	0.16	0.19	0.24
Import Quantity (MMT)	0.07	0.10	0.12	0.11	0.11	0.11	0.11	0.15
Import (Per cent)	4.32	5.10	5.41	5.47	5.97	4.85	4.62	5.05
Domestic Consumption (Rs. Crore)	10,160	10,932	11,448	12,283	12,433	13,104	13,434	13,933
Market Share of Top 10 (Per cent)	30.23	28.34	26.99	25.39	28.42	30.64	36.01	35.21

Source: CMIE, *Industry: Market Size and Market Shares* (2002, 2004).

Table 5.1 provides the key statistics for 239 firms in the paper industry for the period 1995-96 to 2002-03. The top 10 companies control 30-35 per cent of the market. Total production grew from 3.55 million tonnes in 1995-96 to 5.24 million tonnes in 2002-03. Sales grew by 30 per cent in the five-year period from Rs. 10,000 crore in 1995-96 to Rs. 14,000 crore in 2002-03. Imports constitute only around 5-6 per cent of the market. Increasingly, the raw material, namely, pulp is imported due to the limitations placed on clear felling of forests. The data provided for the paper industry do not include newsprint which is imported. Exports which constitute only 2-3 per cent of domestic production, peaked in 2002-03 at 0.24 million tonnes.

Cost Aspects

Pulp, power, caustic soda and chlorine account for the major costs of the industry. While pulp cost accounts for about 30-35 per cent of the production costs, power accounts for about 15-20 per cent. Pulp can be produced from wood, wastepaper, bamboo, and agro residue like bagasse, jute and straw. Based on the data for select firms, it is estimated that the cost of raw materials, power and fuel, together account for over 43 per cent of sales revenue. Other manufacturing costs and interest account for about 18 per cent and 8 per cent respectively of sales revenue (Capitaline, 2002).

Domestic manufacturers are at a disadvantage as the country is short of high quality raw material resources and the use of recycled paper is quite limited. Moreover, the country suffers from cost disadvantage as fibrous raw materials in India cost $ 50 per tonne as against $ 20 in Indonesia and $ 28 in Brazil. The cost of power is also relatively high in India. Further, the industry needs to invest heavily in modernisation to bring in the latest technology to be cost effective. This, however, is difficult as there are a large number of small firms in the industry (Capitaline, 2002). These cost factors have to be kept in mind if technology upgradation has to be undertaken to reduce or eliminate the use of chlorine for bleaching.

Recently, the government has allowed blending of ethanol produced from molasses with petrol up to five per cent. While India currently produces around 1.3 billion litres of ethanol, Brazil with almost the same sugar production capacity produces 14 billion litres. If India expands its production of ethanol from molasses, then the other by-product of the sugar industry, namely, bagasse can be used for the manufacture of paper. Apart from serving as a source of raw material, bagasse requires less chlorine for bleaching.

Central Excise Duty on Paper

Paper, paper board and articles made of paper and of paper board are all covered by tariff entries under Chapter 48 of Central Excise Tariff. The rates of duty of central excise under Chapter 48 have been broad banded. Thus, all dutiable items under Chapter 48 of Central Excise Tariff carry a single rate of duty i.e. 16 per cent irrespective of whether the item is paper or paper board or tissue paper or napkins or printing and writing paper or craft paper or whether it is glazed or coated paper or carbon paper or whether it is an article made thereof.

However, paper and paper board manufactured by use of non-conventional raw materials carry 'nil' duty, provided the pulp used in the manufacture contained more than 75 per cent of the non-conventional raw material. This means that such pulp should not contain more than 25 per cent by weight, materials such as bamboo, hard woods, soft woods, or reeds other than *sarkanda* or rags.

The structure of central excise duties on paper does not take note of whether the chemicals used in the manufacture of paper have any harmful effect on the ecology or environment. One of the salient features of central excise tariff structure, however, is that newsprint, paper for printing of educational textbooks, handmade paper/paper board and braille paper and sanitary articles carry 'nil' duty.

Table 5.2

Chlorine Production and Sales

	1995-96	*1996-97*	*1997-98*	*1998-99*	*1999-00*	*2000-01*	*2001-02*	*2002-03*
Production (in '000 tonnes)	760.3	764.5	717.2	789.6	898	918.3	943.5	1016.2
Sales (Rs. crore)	334.4	334	285.9	276.2	301.5	419.9	284.5	676.4

Source: CMIE, *Industry: Market Size and Shares (2002, 2004).*

Chlorine Production and Use

Chlorine is an intermediate good that has a wide variety of applications including bleaching of pulp and paper, manufacture of polyvinyl chloride, purification of water/wastewater, etc. Production of chlorine has gone up from 0.763 million tonnes in 1995-96 to 1.016 million tonnes in 2002-03. Sales declined in between but have now shot up to Rs. 676 crore in 2002-03 from Rs. 334 crore in 1995-1996 as shown in Table 5.2. Chlorine attracts 16 per cent central excise duty which is rebatable under the CENVAT. Import of chlorine is negligible, since the domestic production is more than adequate. Exports constitute less than 1 per cent. In 2001, only 3339 tonnes out of 9.18 million tonnes of chlorine produced were exported.

Environmental Impact of Chlorine

Paper manufacturers use chlorine or chlorine compounds to bleach the raw wood pulp which is brown in raw form. The use of chlorine in bleaching creates highly toxic chemicals called 'organochlorines' such as dioxins and furans. The organochlorines are released into the environment through air, water, and solid waste discharged by the paper and pulp mills. These toxins accumulate in the food chain as well, and have been associated with a wide range of human health problems.

There are four main bleaching options in the paper-making process:

Elemental Chlorine (EC): The traditional method for paper bleaching uses chlorine gas (elemental chlorine) to whiten paper. This process produces large amounts of dioxins. EC paper can contain either virgin or recycled fibres.

Elemental Chlorine-Free (ECF): Chlorine derivatives such as chlorine-dioxide and hypochlorite are used to achieve reductions in dioxins. ECF paper can contain either virgin or recycled fibres.

Totally Chlorine-Free (TCF): A combination of oxygen delignification is used with hydrogen peroxide or ozone as the bleaching agent. Paper that is made from TCF pulp is made with 100 per cent virgin fibres, including wood and alternative fibres. The TCF label is not allowed to be used on recycled paper because the chlorine content of the original paper is not known.

Processed Chlorine-Free (PCF): PCF contains up to 100 per cent recycled content which has been re-bleached with a process that does not use chlorine compounds. Any virgin fibre in these papers is TCF. Since it is impossible to tell whether their recycled content has been bleached with chlorine in the past, PCF papers cannot be labelled totally chlorine-free.

The global debate over the use of chlorine free bleaching has been largely absent in India. Almost all the pulp and mills continue to use chlorine to bleach pulp. Though the Pollution Control Boards have standards for release of organochlorine compounds, they have started to monitor these pollutants only recently. One method to reduce the quantity of chlorine used is through oxygen delignification before chlorine bleaching, but only some of the larger pulp mills have progressed to this stage. Some mills now use chlorine dioxide in order to partially replace elemental chlorine. Pulp mills in India require a lot of chlorine—about 90 kg of elemental chlorine per tonne of pulp and about 70 kg when chlorine dioxide is used. Chlorine usage in other countries is much lower, around 25 kg per tonne (Edwards and Santillo, 1996).

Although the Central Pollution Control Board has placed a maximum level of two kg of total organic chlorine that would be discharged per tonne of pulp, this standard is unlikely to be achieved since it is often not monitored. Since there are limited testing facilities for organochlorine in most of the state boards, enforcement is virtually absent. The pulp and paper industry has only recently started to take steps to reduce the use of chlorine.

On an average, each paper and pulp plant uses about 3500 MT of chlorine per year (Table 5.3) which works out to 70 kg per tonne of pulp. At a cost of about Rs. 4700 per MT, the annual chlorine cost would be of the order of Rs. 1.65 crore per plant per year or approximately 10 per cent of the cost of the output. Hence, the value added tax (16 per cent) on chlorine would be approximately 1.6 per cent of the output cost in the case of pulp and paper plants, which would normally be rebated under CENVAT.

Table 5.3

Paper and Pulp Production–Use of Chlorine in Selected Firms

	1995-96	*1996-97*	*1997-98*	*3-year Average*	*Average per unit*
Total Production (MT)	1417064.26 (22)	1432736.96 (22)	1486754.62 (21)	1445518.64 (21)	66716.25
Chlorine (MT)	60745.68 (18)	63650.77 (18)	65637.65 (18)	63344.70	3519.15
Elemental Chlorine (Kg/MT)	76.41 (15)	69.27 (16)	65.63 (16)	70.44	70.44
Equivalent Chlorine (Kg/MT)	123.79 (14)	116.21 (15)	119.01 (15)	119.67	119.67

Note: Number in parenthesis indicates the number of units.

Source: Centre for Science and Environment, *Green Rating Project* (2000).

In a Green Rating Project of Pulp and Paper mills by Centre for Science and Environment, it was noted that some of the paper mills were gradually moving away from total reliance on elemental chlorine to the use of other bleaching agents like chlorine dioxide and hypochlorite (Table 5.3) in addition to chlorine (Centre for Science and Environment, 2000).

One firm has undertaken successful trials with enzyme pre-bleaching and has reduced chlorine consumption by 30 per cent. The pilot project being a success the company plans to begin the use of this technology for bagasse pulping. However, the rates of chlorine usage are still very large and there is a high probability that organochlorine compounds in excess of the standard are discharged from most pulp and paper plants. Since total organochlorine is rarely monitored, there is no pressure on the firms to reduce their chlorine consumption. It would, therefore, be desirable to use ecotaxes as a disincentive against the use of chlorine. Since chlorine is an intermediate input one policy measure would be to withdraw the VAT rebate on the excise duty. In effect, this would make the chlorine bleached paper more expensive by the amount of the excise duty on chlorine (i.e., 16 per cent of the cost of chlorine used).

Comments by Industry

The opinion of one of the major paper manufacturers was that a shift to chlorine substitutes was desirable, but that the investment and operating costs were a major deterrent. Paper mills have already started to take some steps to reduce the use of chlorine:

1. Pulp washing systems will reduce the usage of chlorine for bleaching.
2. Similarly, enzymatic prebleaching reduces the usage of chlorine.
3. By purchasing pulp or by using waste paper pulp, the need for bleaching is reduced.
4. Using bagasse as a raw material also reduces the percentage of chlorine used.

Some firms have achieved reduction in the use of chlorine by using other bleaching agents like hydrogen peroxide. For total elimination of chlorine, paper mills will have to install an Oxygen Delignification plant which could cost approximately Rs. 100 crore. This may not be viable for most paper mills with a capacity of 300 tonnes per day. Also, the operating costs will be higher for chlorine free bleaching.

Two suggestions were made by the manufacturer:

(i) Regulate the mills on the basis of the AOX (organo-chlorine standard) emission level.

(ii) Provide an incentive to switch to other types of bleaching.

Regulation of AOX is not feasible at the present time due to lack of adequate testing facilities with most of the State Pollution Control

Boards. The second best solution is to tax the input (chlorine) rather than monitor the emission of AOX. The purpose of not rebating the excise duty on chlorine is to put some pressure on paper mills and viscose rayon plants to switch to chlorine substitutes, since the regulatory approach has not been very successful. Paper mills/viscose rayon plants would have to install additional equipment by way of a chlorine dioxide plant and or an oxygen delignification plant. Accelerated depreciation of 50 per cent of the investment cost is recommended, since it would lead to reduction in the use of chlorine.

Viscose Rayon

Bleaching of pulp is also done for the production of viscose rayon.

Table 5.4

VSF Production and Sales

	1995-96	*1996-97*	*1997-98*	*1998-99*	*1999-2000*	*2000-01*	*2001-02*	*2002-03*
Production (in million tonnes)	0.194	0.179	0.185	0.178	0.202	0.236	0.177	0.225
Sales (Rs. crore)	1485.0	1317.3	1315.2	1291.4	1416.6	1701.9	1412.6	1754.3

Source: CMIE, *Industry: Market Size and Shares* (2002, 2004).

Viscose Staple Fibre (VSF) which is derived from pulp is produced by only a few firms in the country and it is reported that one of the plants has closed down. However, the total production has increased from 0.194 million tonnes in 1995-96 to 0.225 million tones in 2002-03. Sales have grown slowly from Rs. 1485 crore in 1995-96 to Rs. 1754.3 crore in 2002-03. VSF firms have been plagued by pollution problems, particularly the colour from lignin. However, there has been less concern about the organochlorine compounds that are formed due to the use of chlorine. Even if imported pulp is used, it may not be chlorine free unless there are import restrictions on the chlorine content of the pulp.

Table 5.5

Rayon Grade Pulp Production–Usage of Chlorine

	1995-96	*1996-97*	*1997-98*	*Average*	*Average Per Unit*
Total Production (MT)	84375 (3)	114374 (3)	79507 (2)	92752.00	34782.00
Chlorine (MT)	8843.5 (3)	9704.5 (3)	7970.9 (2)	8839.63	3314.86
Elemental Chlorine (Kg/MT)	50.33 (3)	50.07 (3)	45.45 (2)	48.63	48.63
Equivalent Chlorine (Kg/MT)	118.63 (3)	116.83 (3)	122 (2)	119.15	119.15

Note: Number in parenthesis indicates the number of units.

Source: Centre for Science and Environment, *Green Rating Project* (2000).

Recommendations

Pulp and paper mills as well as viscose rayon mills could be encouraged to move towards chlorine substitutes like hydrogen peroxide if the rebate on the central excise duty of chlorine in the VAT system is withdrawn. At the present time, chlorine attracts a basic excise duty (BED) of 16 per cent. The additional excise revenue would be of the order of Rs. 5.3 crore for pulp and paper plants and about Rs. 0.65 crore for viscose rayon plants or a total of about Rs. 6 crore per year. However, the rebate can be continued for chlorine substitutes (even if they contain chlorine) like chlorine dioxide or hypochlorite. This measure would give some incentive to switch over to chlorine substitutes. Some investment in new machinery is called for if a firm wishes to use effective chlorine substitutes. It is recommended that accelerated depreciation (at the rate of 50 per cent) be provided for investment in such machinery, since it will lead to reduction in the use of chlorine.

6

Phosphate Based Detergents

Introduction

There has been a worldwide movement to limit the use of phosphates in detergents, since phosphates are a nutrient source for algae and other organic matter in lakes and water bodies. However, in India apart from detergents, there are also other sources of phosphates such as human/livestock wastes and fertilisers which are quantitatively more significant in terms of the phosphate loads. Some therefore question the need to limit phosphates in detergents. But in India, detergents are used widely by all sections of the population, and the wastewater invariably ends up in water bodies and not in sewage treatment plants. Also, there are thousands of small water bodies such as tanks, ponds, and lakes where washing takes place directly. Drains from settlements also discharge into these water bodies. Eutrophication of these water bodies is already taking place in many parts of the country.

The demand for detergents has been growing rapidly in both urban and rural areas and consequently, the damage to the water bodies from the discharge of phosphate is fast increasing. Therefore, it would be advisable to encourage the use of alternatives to phosphates in the manufacture of detergents. Regulatory steps to limit phosphates can be taken when there is better information on the impact of detergents in Indian conditions. Some Indian brands already have low phosphate content and combine relatively good performance with economy and eco-friendliness, and those brands should be encouraged by reducing the excise duty on them. Manufacturers of these products should also be encouraged to use the eco-label provided by the Ministry of Environment and Forests as a means of promoting their products. These issues are discussed in more detail below.

Detergent Industry

The detergent industry is dominated by two firms that account for nearly 60 per cent of market share. The top 10 firms control 81 per cent of the market. Imports make up less than 3 per cent. Production of detergents has shot up from 0.76 million tonnes in 1995-96 to 2.31 million tonnes in 2000-01. Sales have grown from Rs. 1865 crore in 1995-96 to Rs. 4887 crore in 2000-01.

Table 6.1

Production and Sales in Detergent Industry

	1995-96	*1996-97*	*1997-98*	*1998-99*	*1999-2000*	*2000-2001*
Production (in million tonnes)	0.761	1.018	1.378	1.566	1.910	2.314
Sales (Rs. crore)	1864.6	2513.2	3137.3	3665.9	4210.2	4887.2

Source: CMIE, *Industry: Market Size and Shares* (2002).

The rapid growth of the detergent industry is partly due to the expanding rural market where detergents are now widely used by all classes of the population. This market is likely to grow faster in the years to come.

Use of Phosphates

Most modern detergents use 'builders' to soften the water in order to improve the cleansing action. Phosphates have been the builders most commonly used in detergents. Traditionally, sodium tri-poly phosphate (STPP) was the preferred builder. It provides good water softening, due to its high binding capacity for calcium ions and heavy metals, good buffering properties, which contribute to detergency, and good anti-redeposition properties. STPP was the preferred builder due to its low toxicity (combined with good formulation and in-use characteristics). In the 1970s there was growing concern that phosphates in detergents were contributing to eutrophication (a term used to describe water which is 'over rich' in nutrients), which can lead to an excessive growth of algae. This led to the search for alternative builders, and the development of a synthetic detergent 'Zeolite' which is today the principal alternative to phosphate builders.

Phosphates contribute to an oversupply of nutrients to water bodies. The result is excess algae growth and sometimes these algae may be toxic. Phosphates also contribute to eutrophication of lakes and ponds. The increase in decomposer organisms (that require oxygen) can deplete the oxygen dissolved in the water. Plants and fish can die when the oxygen is depleted. In India, many freshwater bodies, particularly in rural areas, are sources of drinking water supply. Eutrophication of these water bodies would render them useless for water supply. Removal of phosphates from wastewater is possible only when there is treatment at a sewage treatment plant. Only major cities in India treat sewage and often these do not have tertiary treatment required for the removal of nutrients like phosphates. Apart from the impact on aquatic ecosystems, detergents containing phosphates may also cause skin irritation when washing is done by hand rather than in a washing machine.

Sodium Tri-Poly Phosphate (STPP)

Two large firms control the production of STPP in India which is used by the detergent industry. The production of STPP has gone up from 47,148 tonnes in 1995-96 to about 60,000 tonnes in 2002-03. Very little STPP is either imported or exported. Sales revenue rose from Rs. 132.3 crore in 1995-96 to Rs. 208.7 crore in 2002-03. It is not known if the decline in the sales of STPP is due to a decrease in the use of phosphates in detergents.

Table 6.2

Production and Sales of STPP

	1995-96	*1996-97*	*1997-98*	*1998-99*	*1999-00*	*2000-01*	*2001-02*	*2002-03*
Production (in tonnes)	47148	56220	60639	60041	64356	60095	59755	59465
Sales(Rs. crore)	132.3	183.1	182.3	139.3	142.3	128.2	117.7	208.7

Source: CMIE, *Industry: Market Size and Shares* (2002, 2004).

Detergent Use in India

The per capita detergent use in India is 2.8 kg per year, which is likely to go up by about 10 per cent per year since the rural use of detergents is increasing. Detergent manufacturers claim that the phosphate contribution of their products to eutrophication is minimal. According to one estimate, the phosphates from STPP contributing to the phosphate load in the Indian waterways is only 0.23 per cent of the total phosphate content (Driver, 2001). Some experts feel that the removal of phosphate from detergents would not solve the problem of eutrophication. However, long term risks are possible if detergent laden effluents are constantly entering the same inland water bodies. In many rural situations, the same water body is used for large scale washing of clothes, bathing, irrigation and even aquaculture.

Regulatory Approaches

Currently there are no environmental regulations applicable to household detergents. However, the Bureau of Indian Standards has separately laid down standards for ecolabelling of detergents in India. Based on the quality, safety and performance of these detergents, a set of general and specific requirements for an 'Ecomark' have been established by the Ministry of Environment and Forests.

In the case of detergents, there is already a provision for an Ecomark, in the form of an Ecomark plan which has been laid down encouraging phosphate-free detergents, but the level of compliance has not been high according to environmental organisations like Toxics Link. Reductions in the tax on alternative detergents made by using zeolites would help reduce the use of phosphates in detergents. While India's environmental regulations recognise and categorise phosphine, phosphorus and its compounds as toxic chemicals, there are no regulations that are applicable to household detergents.

The Bureau of Indian Standards (BIS) has separately laid down the standards for ecolabelling of detergents in India. Based on the quality, safety and performance of these detergents, a set of general and specific requirements for an 'ecomark' have been established. The specific and general requirements laid down by BIS for ecomarking of detergents state that they should not contain any phosphate. They also stress that the surfactants used in the manufacture of household laundry detergent powders should be readily biodegradable and the products packed in recyclable or biodegradable materials. Although most detergents are yet to comply with the ecomark, many brands have comparatively low level of phosphates.

Comparative test results of 14 brands of popular Indian detergents done by consumer organisation, 'VOICE' showed that the top three contained more than 20 per cent STPP in them by weight (Voice, 2001). Analysts and experts recommend that regulations be put in place immediately that require that no detergent product contains more than 5 per cent phosphorus by weight, and by 2006, no detergent should have more than 2.2 per cent phosphorus, and by 2011, the ratio should be brought down to 0.5 per cent.

Experiences in Other Countries

Countries where phosphates are banned:

- Switzerland
- Norway
- Venezuela
- Japan (Shiga prefecture)
- United States – 12 states

Countries where phosphate use is limited by law:

- Canada 2.2% P
- United States 8.7% P
  (other than 12 states where phosphates are banned)
- Finland 7.0% P
- Japan 2.5% P
- Italy 1.0% P
- Germany 5.0% P (approx.)
- Austria 5.0% P (approx.)

Countries where phosphate use is limited by voluntary agreements:

- Netherlands 5.0% P (approx.)
- Sweden 7.5% P
- France 5.0% P

Source: Veerapan (2001).

Alternatives to Phosphates

Synthetic detergent zeolite is today the principal alternative to phosphate type builders, softening the washing water by calcium ion exchange. Introduced into laundry detergent products in the late 1970s use of zeolite in Europe had reached 50000 tonnes by 1984 and had increased to 600000 tonnes by 1996. Although synthetic detergent zeolites are very effective at 'tying up' or removing calcium ions, they are much less effective in removing magnesium ions. Thus, to obtain optimal water softening, manufacturers usually include a co-builder, often a polycarboxylate, to improve the water softening capacity of synthetic detergent zeolite. In Europe and North America, zeolites have replaced phosphates to the extent of 10–15 per cent. Presently, SPIC produces zeolite and zeolite based detergent. Henkel SPIC has built a 10,000 tonnes per annum (tpa) plant for zeolite manufacture. A market survey showed that 25-30 per cent of the detergent market may shift to zeolite based detergents in the next 5-6 years, and there is likely to be a large supply gap for zeolite.

Human risk assessment studies have shown that the use of zeolite A (sodium aluminium silicate) in household detergents does not cause concern with regard to consumer use (HERA, 2002). zeolite manufacturers also claim that there could be cost savings to the tune of Rs. 1500 per tonne of detergent if STPP is replaced by zeolite supported by a proper co-builder. It would therefore be desirable to promote the use of non-phosphate builders like zeolite through appropriate tax policies.

Central Excise Duty

All synthetic detergents attract a central excise duty of 16 per cent. The excise revenue collection for both soaps and detergents for the year 2001-02 was around Rs. 700 crore. Since sodium tri-poly phosphate (STPP) is an intermediate compound in the production of detergent, the duty payable on STPP would be rebated in the present CENVAT system.

Comments by Detergent Manufacturers

The Indian Soaps and Toiletries Makers' Association visited Madras School of Economics on November 4, 2003 and made a presentation. Subsequently, they also sent their comments in a letter on the proposals for an ecotax on detergents based on phosphate content. Their major observations are as follows:

1. STPP should not be singled out for levying an ecotax since phosphates in detergents contribute only minimally to the phosphate load. Also, there are few studies on eutrophication in Indian waterways and the linkage to STPP.
2. The Association states that it is not seeking any incentive for a specific ingredient such as zeolite or sodium carbonate but rather looking for innovative technologies that would reduce environmental impact.
3. The Association proposes a reduction in excise duty from 16 per cent to 8 per cent for compact detergents since these reduce chemical load, packaging, and energy consumption.
4. The Association also recommends waiving of import duties on washing machines using innovative technologies based on the detergentless principle. These technologies will produce a very small quantity of waste. Detergentless technologies will result in savings in water and will also decrease the discharge of effluents. These technologies will be of benefit to hotels and other institutions which use large volumes of water.

In general, the Association is in favour of economic instruments for environmental protection, but does not recommend taxes/ incentives based on phosphate content.

On the other hand, one of the manufacturers, Henkel SPIC agreed with the ecotax report that the use of non-phosphate detergents should be promoted. They argue that 'zeolite' based detergent products are a cost effective alternative to STPP based detergents. zeolite is non-toxic and does not contribute to eutrophication of water sources. It has other favourable characteristics like antiredeposition, anticaking, chemical stability and optical brightness.

Recommendations

1. In line with the precautionary principle, it is desirable to limit the use of phosphates in detergents, even though there is not much information on the impact of phosphate based detergents in India. The major input in detergents which contributes to the phosphate content is sodium tri-poly phosphate (STPP). The rebate of the CENVAT levied on the

phosphate compounds used as inputs (such as STPP) can be denied. The production and domestic consumption of STPP in India in 2002-03 was about 60,000 tonnes with a value of approximately Rs. 200 crore. If the sales value of STPP is approximately Rs. 200 crore, the excise duty levied at16 per cent would be Rs. 32 crore. If the CENVAT on STPP is not rebated, there would be additional revenue to the Government of Rs. 32 crore.

2. Non-phosphate detergents could be promoted by decreasing the excise duty from 16 per cent to 8 per cent which would be revenue negative. The production of detergents is about 200,000 tonnes. Zeolite now accounts for about 10,000-15,000 tonnes. Ultimately, low phosphate detergents may account for 10-15 per cent of the market. According to the Ministry of Finance, the actual excise revenue for detergents for 2002-03 was Rs. 561 crore, and the assessed value was approximately Rs. 3520 crore. Low phosphate detergents may account for only about 10 per cent. If the excise duty is reduced from 16 per cent to 8 per cent, the loss in revenue would be about Rs. 28 crores.

3. Two other recommendations by the ISTM Association are as follows:
   a. The Indian Soaps and Toiletries Makers' (ISTM) Association has recommended a reduction in excise duty from 16 per cent to 8 per cent for compact detergents since these reduce chemical load, packaging and energy consumption.
   b. The ISTM Association also recommends reducing import duties on large size washing machines incorporating innovative technologies based on the detergentless principle. Detergentless technologies will result in savings in water and will also decrease the discharge of effluents. These technologies will be of benefit to hotels and other institutions which use large volumes of water.

The revenue impact of these two recommendations would be minimal with positive environmental impact.

7

Chemical Pesticides and Biopesticides

Introduction

Crop protection chemicals are widely used in India since the value of the crop loss due to pests is estimated to exceed Rs. 50 billion. Chemical pesticides, particularly insecticides, are used to control pests in crops like cotton, rice, pulses, fruits and vegetables. However, spraying of pesticides results in residues in drinking water, vegetables, milk, fish, etc. There is also occupational exposure of farm workers who apply the pesticides. Many of the pesticides used are hazardous to human and animal health, and some have been banned. Agricultural scientists have recommended the use of integrated pest management practices and the use of biological control of pests. Biopesticides such as *neem*-based pesticide have been promoted in India and abroad. Biopesticides are inherently less harmful than chemical pesticides and are also target specific affecting only the target pests. However, most biopesticides are generally more expensive than chemical pesticides when calculated on a dose per hectare basis. Hence, the government should promote their use by removing the central excise duty on biopesticides. Such a policy would be similar to the preferential treatment given to organic fertilisers. Any loss of revenue would be minimal compared to the environmental damage caused by chemical pesticides.

Pesticide Production in India

India is the largest producer of chemical pesticides in South Asia with a production of 88,751 MT in 1998-99. It is the second largest manufacturer of basic pesticide chemicals in Asia next to China, and ranks 12th globally. The total installed capacity is about 1,24,000 MT for the manufacture of 62 technical grade pesticides which meet 95 per cent of the total needs of the country.

Table 7.1

Production and Consumption of Technical Grade Pesticides in Agriculture

Years	*Production (MT)*	*Consumption (MT)*
1994-1995	90758	80000
1995-1996	96880	73652
1996-1997	94350	66677
1997-1998	84154	60143
1998-1999	88751	57240
1999-2000	Not available	54135

Source: Directorate of Plant Protection, Quarantine and Storage, Ministry of Agriculture & Cooperation, Government of India, Faridabad.

Both China and India are substantial suppliers in the generic market with major exports. In 2001-02 when domestic sales were nearly Rs. 4000 crore, exports were expected to rise to Rs. 1800 crore.

Pesticide Use in India

India is the third largest consumer of pesticides in the world and the highest among the South Asian countries. During the last four decades, India has increased the consumption of pesticides several hundred fold, from 154 MT in 1953-54 to 80,000 MT in 1994-95. However, thereafter the consumption of pesticides declined to the present level of 54,135 MT in 1999-2000 (Table 7.1). The decline was primarily due to the ban or restriction on the use of organochlorines, which have high application rates, and the introduction of Integrated Pest Management programmes.

In spite of the significant increase in pesticide consumption, India with about 4 per cent of the world cropped area has a share of only around 1.7 per cent of world pesticide consumption. In contrast to the global scenario, the Indian market is dominated by insecticides (76 per cent) followed by fungicides (13 per cent) and herbicides (10 per cent). The low percentage for herbicides reflects the availability of cheap labour for weeding operations.

The area under plant protection is continuously increasing, but still only about 25-30 per cent of the total cultivated area is under pesticide cover. The average per hectare consumption of pesticides in India in agriculture (calculated on the basis of total consumption of technical grade pesticides divided by the gross cultivated area) was 1.2 g/ha in 1953-54 which increased to 377 g/ha 1985-86 and to 431 g/ha in 1992-93. However, thereafter consumption gradually declined to 288 g/ha in 1999-2000. The values are much lower compared to the pesticide consumption level in the other countries. As the vast majority of rainfed crops do not consume any quantity of pesticides, the average values are very low. But crops like cotton, rice and vegetables consume excessive quantities.

Currently, pesticide use is highly concentrated in a few crops. Cotton consumes about 44.5 per cent of the total value, followed by rice which accounts for about 23 per cent. These two crops alone consume more than two-thirds of the total quantity of pesticides in the country. There is injudicious application of pesticides by farmers without proper guidance from agricultural extension experts. Consequently, there has been development of resistance to pesticides and pesticide residue is present at high levels in many agricultural and dairy products. Cropping systems which were naturally more pest resistant have also been given up by farmers due to the excessive reliance on chemical pesticides.

Pesticide Resistance

The major requirement for any sustainable production of crops is to guard them against the onslaught of pests, diseases and weeds without harming the natural predators of pests and causing harm to the environment. Scientists throughout the world have been addressing the problems of long term ill effects of chemical pesticides used in agriculture. Pesticides are known polluters of environment, and their improper use has caused damage. The negative externalities associated with pesticide use are:

- pesticide resistance,
- pest resurgence,
- public health hazard, and
- environmental and food contamination.

Excessive and indiscriminate use of pesticides has led to the development of secondary pests; the destruction of the natural predators of pests also contributes to pesticide resistance and pest resurgence.

Toxic Residues in Food

Pesticide residue in vegetables and fruits is a major health hazard. A study of farmgate vegetables showed that 55 per cent of 4111 samples were contaminated with pesticide residue and 9.5 per cent of the samples had more than the maximum recommended level (Agnihotri, 1999). Pesticide residues have also been found in the drinking water in some states and also in soft drinks. Pesticides adversely affect animals as well as humans. Chronic exposure to a low concentration of pesticides induces the detoxifying enzymes in the liver. Some organo-pesticides affect the nervous system, impairing neural transmission. Impaired metabolism leads to the decline of animal population.

Pesticide education, monitoring and use of safety equipment are largely non-existent in the country. Hence, the excessive and improper use of pesticides has very serious consequences. Pesticides affect the general public through the food chain. Pesticides in the diet get deposited in the fat tissue. The levels of DDT and its derivatives in human milk causes an increased level of this pesticide in infants. Developed nations are switching over to organic farming with bio-control measures to produce nutritious food, while in India indiscriminate use of pesticides creates health hazards to human beings and animals. There is need to undertake long term toxicological studies, especially in view of the persistent association and incrimination of pesticides with cancer, infertility, and birth defects. There is increasing concern about pesticide levels in bottled water and soft drinks (Centre for Science and Environment, 2004).

Occupational Exposure

Most exposed workers are agricultural labourers and small farmers in areas of plantation or intensive agriculture. Inadequate safety measures and unhygienic practices are very common in applying, storing, transporting, and manufacturing pesticides. Most farm workers are not trained in safe pesticide use and the few regulations that address farm worker safety are unrealistic or not enforced.

Health and safety issues are exacerbated by a general lack of hazard awareness. The lack of protective clothing or difficulty in wearing protective clothing in tropical climates, shortage of facilities for washing after use or in case of accidents, the value of containers for re-use in storing food and drink, illiteracy, labelling difficulties relating either to language complexity or misleading information, lack of regulatory authorities and lack of enforcement lead to high exposure of farm workers to pesticides.

Workers are exposed usually through the skin when using improper equipment to mix or spray pesticides. Organophosphates and carbamates are particularly capable of absorption through the skin even if cotton clothing is worn. Kitchen spoons, matchboxes, tin cans and bottles are often used to mix active ingredients which allows highly concentrated chemicals to touch bare skin.

Registration and Licensing of Pesticides

In India, pesticides are compulsorily registered before their introduction for use in the country by the Registration Committee which has been constituted under Section 5 of the Insecticides Act 1968. The Registration Committee registers pesticides after satisfying itself regarding the safety of the pesticides for human beings, animals and the environment. The government reviews the use of pesticides from time to time by appointing an Expert Committee and decides on the recommendations of the Expert Committee and/or Registration Committee. The use of chemical pesticides in the recommended dose with the prescribed method of application at the prescribed intervals does not cause any adverse effect on crop production in the country. To check the supply of substandard/mis-branded pesticides, provisions exist in the Insecticides Act 1968. The central and state government officers are authorised to take action against the manufacturers/ suppliers, if the samples are found to be substandard/misbranded. In order to counter the ill effects of chemical pesticides, the Government of India has already banned 23 hazardous pesticides/pesticide formulations and restricted the use of another 10 pesticides.

All pesticide handlers have to be licenced by the government. Handlers include importers manufacturers, formulators, exporters, re-packers, distributors, warehouse personnel and sales people. A separate licence is required for each establishment of place of business subject to these rules, and it must be conspicuously displayed on the premises. Government licence permits are needed to sell pesticides at a particular price. However, only half the pesticide dealers hold permits, because the application and renewal processes are quite cumbersome.

Pricing of Pesticides

Retail prices of pesticides are significantly affected by taxation, tariffs and the exchange rate. An effective pricing policy might apply an ecocess, taxing less hazardous pesticides at lower rates than more hazardous ones. However, given the toxic nature of pesticides, a direct ban of the most hazardous pesticides would be preferable to discriminatory taxation.

The rate of central excise duty on all varieties of pesticides including biopesticides is 16 per cent. There is no preferential treatment of biopesticides.

The Biopesticide Market

Rough estimates by All India Biotech Association (AIBA) indicate that biopesticides constitute less than one per cent of the crop protection market, consisting of a total quantity of about 3000 tonnes. In 1999-2000, around 740 MT of *neem*-based pesticides and nearly 135 MT of *Bacillus thuringiensis* (Bt) biopesticides were used under the IPM programme of the Union Ministry of Agriculture. While most of the biopesticides are used in India, *neem* is extensively marketed outside the IPM programme, especially in the tea estates of West Bengal, Assam and South India, as well as for the protection of vegetable crops. Nevertheless, the total area under biopesticides in India was only around 0.1 million hectares in 1999-2000—a fraction of the total cropped area. There is however a significant export market for biopesticides, especially *neem*-based formulations, which are exported to North America, Europe, Japan and East Asian Countries.

Unlike the situation in the world market where bacterial pesticides form the major component of biopesticides, in India plant extracts, especially *neem*-based botanical pesticides, form the major sales component, followed by bacterial pesticides such as Bt, viruses, pheromones and biocontrol agents like fungi and predator insects. *Neem*-based pesticides, Bt and certain pheromones are the only biopesticides that are produced at the industrial level. Most of the other biopesticides are used only for extension activities by various universities.

There are three major factors which affect the growth of biopesticides in India. Lack of formal extension work and awareness creation are indeed the most significant aspects. Although, biopesticides are effective on a long term basis, unlike the chemical pesticides, they do not cause instant death of the target organisms. Nevertheless, during the crop duration and over the long term, biopesticides are generally effective not only in controlling pests but also in abating harmful environmental impact. The second important factor affecting their growth is the dominance of the chemical pesticide industry which controls not only the worldwide market but also that in India. Third, farmers depend heavily on the manufacturers and dealers who generally promote the sale of chemical pesticides. This is compounded by the comparatively high cost of biopesticides, apart from the relatively short shelf life especially for the so-called live biopesticides such as the biocontrol agents (fungi, bacteria, viruses, predators) (Alam, 2000).

Botanical Pesticides

Neem has been a subject of intensive international investigation during the past three decades. All parts of the tree are useful and its derivatives find application in diverse areas including pesticides and allied agro-chemicals, human and veterinary medicine, toiletries and cosmetics, fertiliser nitrogen regulation, energy and others. The application in pest control in particular has received global attention. Of the major applications of *neem* in agriculture, standard products based on *azadirachtin-A* have been developed and are commercially available. The technologies to make *azadirachtin-A/meliacin* concentrates and other emulsifiable concentrates based on oil as well as other extracts have been developed.

The dominant position of *neem* botanical pesticides in the biopesticide industry in India is evident in the involvement of large companies in this activity. In contrast to *neem*-based pesticides, other biopesticides have not attained commercial level production in India. Even in the case of the bacterial pesticide Bt, only one firm has commercial level production with a capacity for production of 100 tons of Bt and 20 tonnes *Bacillus sphaericus* for mosquito control. However, domestic production is supplemented by imports, which are less expensive. *Neem*-based pesticide, being the largest segment in the Indian market, is likely to have a combined sale of 15 to 20 lakh litres currently with an annual turnover of Rs. 40 to 45 crores.

Botanical pest control is a distinct possibility in agriculture, since we are endowed with the biodiversity of such plants. More than 1000 species of plants are known to have insecticidal properties, 380 species anti-feedant properties, 300 species repellent properties and 30 species each possessing attractant and insect growth regulator (IGR) properties.

Biopesticides are at present generally more expensive than chemical pesticides e.g. *neem* extracts enriched in the active ingredient *azadirachtin* to the strength of 1500 PPM have a maximum retail price of Rs. 250 to 300 per litre, substantially higher than common chemical pesticides when calculated on dose-per-hectare basis. Similarly, 1 kg of Bt sells in the retail market for Rs. 1000 to 2000, much higher than the cost of chemical pesticides having similar target pests. The long term effectiveness and lack of environmental hazards make biopesticides preferable. As such, enabling taxation and related government policy framework should make biopesticides more competitive.

Acceptance of Biopesticides by Farmers

A study of biopesticides in Haryana indicated that integrated pest management was practiced only by a small percentage of farmers there. The reasons for low acceptance by farmers include, lack of awareness, lack of skills, lack of faith in IPM and the factor that IPM practices are difficult and cumbersome. The efforts of government agencies to popularise biopesticides and IPM have had little impact. In the absence of active promotion by the agriculture departments, the demand for these products has not developed (Alam, 2000).

Recommendations

1. If chemical pesticides are to be discouraged, then an effective pricing policy would be to levy an ecotax on chemical pesticides based on the toxicity level. The most hazardous pesticides are already banned in India. The tax should be a separate ecocess which could be levied on the pesticide manufacturers. The yield from the cess could be used to promote the use of biopesticides and integrated pest management.

   If we assume that the value of domestic consumption of chemical pesticides is roughly Rs. 6000 crores, the expected excise revenue would be Rs. 960 crore since chemical pesticides attract excise duty of 16 per cent. According to the

Ministry of Finance, the actual excise revenue from chemical pesticides for 2002-03 was Rs. 766.83 crore while the assessed value was Rs. 5190 crores.

An additional cess based on toxicity of chemical pesticides, could be collected as an ecotax. The Central Insecticides Board has designated a colour coding based on toxicity. This code could be used as a basis for levying the ecocess.

Toxicity	*Colour*	*Cess*
Extremely Toxic	Bright Red	8%
Highly Toxic	Bright Yellow	6%
Moderately Toxic	Bright Blue	4%
Slightly Toxic	Bright Green	2%
	Average	5%

Source: Central Insecticides Board & Registration Committee.

The estimated additional revenue from the ecocess on chemical pesticides: Rs. 6000 crore at 5 per cent would be Rs. 300 crore.

2. The rate of excise duty on all varieties of pesticides including biopesticides is 16 per cent. To promote the use of biopesticides, the excise duty could be reduced to 8 per cent. The revenue implication would be small since the biopesticides account for 1-2 per cent of the pesticide market.

   If we assume that the consumption of biopesticides reaches 1000 MT, they would account for only 2 per cent of the pesticide market of 50,000 MT. We do not have the sales data, but if we assume that biopesticides account for 1–2 per cent of the pesticide market then the excise revenue from biopesticides would be approximately Rs. 8-16 crores. If the excise duty is reduced from 16 per cent to 8 per cent, the decrease in revenue would be about Rs. 4-8 crores.

8

Fertilisers

Introduction

Chemical fertilisers are a crucial source of the nutrients required for crop growth and food production, and have contributed significantly to the increase in food production in India and elsewhere. The high yielding potential of improved seeds requires augmenting the nutrients in the soil through the application of fertilisers. However, excessive and/or inefficient use of chemical fertilisers leads to environmental damage such as ground water pollution and long-run adverse impact on soil fertility. In India, fertiliser application is generally not based on sound agronomic principles. More importantly, the policy of subsidising chemical fertilisers, followed in India since the early 1970s, has promoted excessive and indiscriminate use of fertilisers. Although, the extent of fertiliser use in India is considerably less than in western countries, the unbalanced and indiscriminate use of chemical fertilisers has led (and is leading) to considerable environmental damage. Some western countries have imposed taxes on fertilisers as a means of limiting the use of chemical fertilisers (See Chapter 2). There have been attempts to promote the use of biofertilisers but as yet they have not received wide acceptance because the quality is not good enough.

In this chapter we discuss (a) the pricing policies followed in regard to chemical fertilisers and the effects of those policies on fertiliser use, (b) policies for the promotion of the use of biofertilisers and (c) the desirability and feasibility of imposing an ecotax on chemical fertilisers.

Fertiliser Use in India

India is the third largest producer and consumer of chemical fertilisers in the world, and accounts for 12 per cent of world consumption. China and India alone account for more than half of the world's urea consumption. The use of chemical fertilisers went up rapidly after the Green Revolution and has also led to a corresponding increase in the production of foodgrains (Table 8.1). However, in the last five years, fertiliser consumption has tended to stabilise around 17 million tonnes.

Table 8.1

Trends in Fertiliser Consumption and Foodgrain Production

Year	Fertiliser Consumption (MT)	Foodgrain Production (MT)	Index of Agricultural Production 1981-82=100
1951-1952	0.05	51.99	46.9
1961-1962	0.34	82.71	68.1
1971-1972	2.66	105.17	85.2
1981-1982	6.07	133.30	109.2
1991-1992	12.73	168.37	145.5
2000-2001	16.70	196.07	178.1*

Note: * for 1998-99.
Source: *http://www.iffco.nic.in*

At present, there are 65 large sized fertiliser plants in the country manufacturing a wide range of nitrogenous, phosphatic and complex fertilisers. There are about 80 medium and small scale units which produce single super phosphate. The gap between demand and supply is met through imports. Imports of urea have declined substantially during the last five years. The country has achieved near self sufficiency in urea production, while 85 per cent of the phosphate requirement is produced by domestic firms (Table 8.2). For potash however, India is dependent on imports, as it has no known reserves.

Fertiliser Consumption Per Hectare

Fertiliser consumption per hectare has grown more than five fold from 16 kg/ha in 1971-72 to 90.12 kg/ha in 2001-02 (Table 8.3). But this is lower than the fertiliser consumption in other South Asian countries like Pakistan and Bangladesh.

Table 8.2

Trends in Production, Consumption and Imports

Year	Urea (N) (in lakh MT)				Diammonium Phosphate (P) (in lakh MT)			
	Production	Consumption	Imports	Imports as a Per Cent of Consumption	Production	Consumption	Imports	Imports as a Per Cent of Consumption
1991-1992	128.31	140.04	3.91	2.8	28.65	45.18	20.77	45.97
1992-1993	131.23	149.06	18.57	12.5	25.95	40.52	14.51	35.81
1993-1994	131.48	158.10	27.83	17.6	19.68	34.80	15.69	45.09
1994-1995	142.83	171.12	28.70	16.8	28.23	35.86	8.65	24.12
1995-1996	158.20	179.69	37.82	21.1	26.47	34.51	15.14	43.87
1996-1997	156.20	190.25	23.28	12.2	27.59	36.24	5.34	14.74
1997-1998	185.96	196.16	23.89	12.2	36.91	53.72	14.6	27.18
1998-1999	192.91	203.96	5.57	2.7	38.68	58.28	21.05	36.12
1999-2000	198.30	203.14	5.33	2.6	38.63	69.38	32.68	47.10
2000-2001	196.47	191.69	0.00	0.0	48.88	57.12	8.60	15.05

Source: Ministry of Chemicals and Fertilisers, *Draft Report of the Working Group on Fertilisers for the Tenth Plan* (not dated).

Table 8.3

Fertiliser Consumption Per Hectare

Year	*Kg/ha*
1971-1972	16.08
1981-1982	34.33
1991-1992	69.84
2001-2002	90.12

Source: *CMIE* (2001).

Within India, there is a lot of variation in the extent of fertiliser use from a low of 2.4 kg./ha in Arunachal Pradesh to 163.27 kg/ha in Punjab. Among the large states, West Bengal, Haryana, Punjab, Uttar Pradesh, Andhra Pradesh, Karnataka, and Tamil Nadu use more than 100 kg/ha of fertiliser. The N:P:K ratio also varies widely among the states. Fertiliser use is higher per hectare in irrigated area. Large inter-farm variations in fertilisers use are also reported in farm management studies. Horticultural crops which are growing in importance are heavy users of fertilisers.

Fertiliser Pricing Policy

In the mid-1960s when the Green Revolution began, the government took a number of steps to promote the establishment of indigenous fertiliser plants. Originally, a central fertiliser pool was created and the fertiliser was sold at a uniform price set equal to the average of cost of domestic production and of imports, with no profit or loss. This system operated only till 1973 when the increasing oil prices led to a substantial increase in production costs (Venkateshwarulu and Sen, 2002). It was then considered necessary to subsidise the price of fertilisers. This meant that the prices of fertilisers were kept well below cost for the farmers but the producers were ensured an adequate return on the capital.[1] The difference was the fertiliser subsidy. All three categories of fertilisers–urea, phosphate

1. They were paid the difference between their estimated cost and the price at which they had to sell.

and potash—were covered by the subsidy scheme till 1992-93. In August 1992, the prices of phosphate and potash were decontrolled, but urea continued to be subsidised as before. After some time, the other two categories were brought under a concession scheme (subsidy).

Subsidising fertiliser prices in a large way led to three major consequences: (1) It led to excessive and indiscriminate use of chemical fertilisers. (2) The burden of subsidy on the Government budget rapidly mounted (see Table 8.4). (3) Inefficient producers were kept alive and as such there was waste of resources. Because of these large adverse effects, the government has decided to rationalise fertiliser pricing on the basis of the recommendations of the Expenditure Reforms Commission.

Table 8.4

Expenditure on Subsidy/Concession

(in Rs. crore)

Year	*Urea*	*Decontrolled Fertilisers*	*Total*
1980-1981	505.26	-	505
1990-1991	4389.06	-	4389.06
1995-1996	6235.00	500.00	6735.00
1996-1997	5906.08	1671.77	7577.85
1997-1998	7321.96	2596.00	9917.96
1999-2000	8744.07	4500.00	13244.07
2000-2001	9481.00	4319.00	13800.00

Source: Ministry of Chemicals and Fertilisers, *Draft Report of the Working Group for the Tenth Plan* (not dated).

While subsidising all fertilisers would encourage over-use, subsidising only urea would lead to unbalanced use, as it happened after the decontrol of phosphatic and potassic fertilisers. After the subsidy was restored to them in some form, the N:P:K ratio improved to 7:2.7:1 in 2000-01, but that is still higher than the ideal of 4:2:1 for Indian conditions. It is known that any policy that encourages the use of only one type of fertiliser, such as nitrogen, will lead to decline in productivity and reduction of response to application of nitrogen (Awasthi, not dated).

Problems to be Considered

On the basis of the foregoing discussion, we may conclude that giving any incentive leading to the over-use of any type of fertiliser (particularly, nitrogen) should be strictly avoided. This is to prevent unbalanced use. But it is also necessary to give up completely or largely the policy of subsidising the use of chemical fertilisers not just because of the subsidy burden, but because there is actually need to reduce the excessive use of chemical fertilisers, if that is possible without significantly affecting output.

Nitrate Pollution

The harmful environmental effects of the use of nitrogenous fertilisers are well known. There are losses of nutrients through leaching, volatilisation, and runoff. 50-60 per cent of applied fertiliser nutrients are lost to the environment. Excessive use of nitrogen cannot be absorbed by the crop and leads to poor recovery, which is also a waste of resources. The nitrogen in the fertiliser is converted to nitrate which is very soluble and can leach into the soil along with irrigation or rain water or carried away by runoff (Bhatnagar and Sharma, 2002).

There is increasing evidence of nitrate pollution of groundwater in states with high fertiliser use like Punjab, Delhi, Maharashtra and Andhra Pradesh. At many locations, nitrate concentration in the groundwater has exceeded the permissible limit of 45 mg/litre (Rao and Puttanna, 2000). Measures to monitor and limit environmental impact are required to avoid increasing damage. Nitrates have been implicated in a wide variety of human health problems including cancer as well as being toxic to animals. The most well-known health problem caused by nitrates is methemoglobinemia in babies. Since technologies available for decontaminating nitrates in drinking water are very expensive, it would be preferable to minimise nitrate leaching from agricultural activities.

Biofertilisers

Most experts feel that biofertilisers can be used to supplement the use of chemical fertilisers. When only synthetic chemicals are added, the soil gradually loses its organic matter and microbiotic activity. Biofertilisers nourish the beneficial micro-organisms in the soil, and greatly increase the soil's humus content. The disadvantage of biofertilisers is that the plant cannot utilise the nutrients immediately since they are released slowly. Hence, it is also difficult to be precise in meeting the crop's nutritional needs. The integrated plant nutrient system (IPNS) which is a combination of chemical and biofertilisers is considered to be the best method of supplying plant nutrition and increasing the crop yield.

Biofertilisers known as microbial inoculants are artificially multiplied cultures of certain soil organisms that can improve soil fertility and crop productivity. Commercial exploitation of such biological processes has occurred only recently. Biofertilisers fix atomospheric nitrogen or solubilise the insoluble phosphates into soluble form for uptake by the crop. The major biofertilisers are *Rhizobium* (for pulses and leguminous oilseeds), *Azotobacter* (for cereals and sugarcane), *Azospirillum* (for smaller millets), blue green algae (for rice) and phosphate solubilising organisms (for all crops). Biofertilisers can provide 15-40 kg of nitrogen per hectare and around 25 kg of phosphate per hectare and supplement up to 25 per cent of the nutrient requirements (Ministry of Chemicals and Fertilisers, not dated). Besides accessing nutrients, different biofertilisers provide growth promoting factors to plants. By controlling soil borne diseases and improving the soil health and properties, these organisms help not only in saving, but also in effectively using chemical fertilisers and result in higher yield rates.

A central sector scheme, National Project on Development and Use of Biofertilisers, has been in operation since the end of the Sixth Plan. The main components of the scheme included:

- Production and distribution of biofertilisers (discontinued since October 2000).
- Training courses for extension workers/agriculture officers.
- Central financial assistance up to Rs. 20 lakh for setting up units of 150 metric tonnes capacity (scheme has been discontinued).
- Field demonstrations/fairs for creating awareness.
- Quality control.

Thus, two important components of the scheme are not in operation now. The estimated potential demand for biofertilisers is around 66,500 MT based on application in 25 per cent of the gross cropped area. The estimated production has increased from about 1000 MT in 1990-91 to 7,390 MT in 2001-02. There are 81 Government of India assisted units with a capacity of 10,300 MT and 39 private units with a capacity of 8000 MT. The bulk of the growth of the biofertiliser industry took place between 1992-1995 and has stagnated thereafter (Table 8.5). Some of the reasons for the low demand for biofertilisers in the short run are short shelf life, lack of quality control, poor storage facilities, lack of awareness and inconsistent crop response.

Table 8.5

All India Capacity and Production of Biofertilisers

(in tonnes)

Year	*Capacity*	*Production*
1992-93	5400	2005
1993-94	6125	3084
1994-95	8114	5800
1995-96	10680	6692
1996-97	12647	7407
1997-98	NA	7105
1998-99	16446	5972
2001-02	15439	7390

Source: The Fertiliser Association of India (2003).

Factors Favouring and Limiting the Use of Biofertilisers

Like other products of biotechnology, biofertilisers are produced from renewable resources, unlike chemical fertilisers which require petroleum based feed stock such as naphtha or natural gas. The cost of biofertilisers is quite low compared to chemical fertilisers since their production uses much less energy. There is also less contribution to green house gas emissions as well as less pollution in production. Most important, there are gains at the farm level from lower surface and groundwater pollution. Biofertilisers contribute to soil health while chemical fertilisers reduce soil fertility. However, biofertilisers cannot be substituted for chemical fertilisers. They can only supplement the use of chemical fertilisers in meeting total nutrient requirement under the integrated plant nutrient management system (IPNS).

Some experts feel that the low acceptance of biofertilisers by farmers is due to their unreliable quality. The present policy of providing grants for biofertiliser production has resulted in a large number of inefficient plants which do not produce good quality biofertilisers. This policy needs to be reviewed and changed. Greater research and development efforts are needed to develop biofertilisers which are easier to store and apply. Production technology and quality control methods have to be improved (Alam, 2000). Quality control of biofertilisers is crucial for obtaining effective response to the field crops. With the assistance of the Bureau of Indian Standards, the specifications for the four major biofertilisers have been notified for ensuring their quality check by state governments.

The farmer has to incur considerable initial cost in terms of skill acquisition, trial and failure, and risk. Adoption of new practices may take time to yield results. Producer firms are said to face serious uncertainty about the demand for the product which deters investment (Ghosh, 2001). Public intervention is justified for building up a market for an input which would have long term benefits. The government has not been proactive to popularise the use of biofertilisers and has withdrawn the central financial assistance scheme. Some of the states have been providing subsidies, up to 50 per cent. For example, Tamil Nadu provides packets of biofertilisers at 25 per cent subsidy to encourage their use by farmers.

Fertiliser Taxes in Other Countries

European countries like Norway, Sweden, and Denmark have introduced fertiliser taxes. In some countries, a nitrogen based levy is made on the basis of a nitrogen account i.e. the net-balance of nitrogen discharged after uptake by crops. In Sweden, it is estimated that the fertiliser tax resulted in reduced demand for fertilisers by 15-20 per cent and also reduced financially optimal dosages by about 10 per cent. Nitrogen based taxes in Europe are levied not only to decrease chemical fertiliser use but also the use of animal manure. There is some debate on whether the proceeds of the nitrogen tax should be recycled. Revenue from the tax could be used to clean up the groundwater.

The purpose of the nitrogen based levy is not only to restrict the pollution of ground and surface water by nitrates, but also restrict greenhouse gas emissions of nitrous oxide (N_2O) which can be released by the use of nitrogenous fertilisers.

Comments by the Fertiliser Association of India

The Fertiliser Association of India (FAI) has expressed its views in a note dated January 30, 2004. The note indicates that fertiliser consumption in India needs to be increased to double foodgrain production in the next 10 years. The average use of fertiliser of 85 kg/ha is lower than in many developed and neighbouring countries. FAI contests the view that there is pollution of the groundwater due to leaching of nitrate, since the average quantum of fertiliser use is low. FAI argues that the loss of soil fertility is caused by inadequate and imbalanced use of fertiliser nutrients, and there is no justification for discouraging the use of chemical fertilisers by imposing a tax. FAI feels that the use of biofertilisers needs to be encouraged to supplement the use of chemical fertiliser in meeting the total nutrient requirement under the integrated plant nutrient management system.

While it is true that the average consumption of fertiliser in India is not very high, the rate of fertiliser application is high in certain areas of the country with intensive agriculture. It has also been pointed out that the efficiency of fertiliser use is a matter of concern (Shrotriya, nd). Nitrogen use efficiency in rice crop is only 30 to 35 per cent, with an overall efficiency level of 50 per cent. The excess fertiliser is likely to leach into groundwater, or run off into surface water. An article in *Current Science* reports increasing evidence of nitrate pollution of groundwater in Punjab, Delhi, Maharashtra and Andhra Pradesh and even in semi-arid areas like the Deccan Plateau and Rajasthan (Rao and Puttanna, 2000). It is important to ensure that the application of nitrogen by farmers is not excessive.

We agree with FAI on the need for increasing production of chemical fertilisers. However, the excessive use of urea in certain irrigated areas, prompted by under-pricing of urea, should be curbed. This will also ensure that there is balanced application of fertilisers. The removal of subsidy for urea will act as a sufficient deterrent against the over-use of urea. A separate ecotax on fertilisers may be considered only in the future. We agree with the FAI that biofertilisers may be promoted as a supplement and not as a substitute for chemical fertilisers.

Recommendations

If there is complete decontrol of the price of urea by 2006 as had been planned, then there will be no inducement to increase the use of chemical fertilisers. If however, the subsidy on urea is not removed, the concessions on the other fertilisers should be retained to ensure balanced application of nutrients. The consumption of chemical fertilisers had reached about 16 million MT in 2000-01 and is estimated to reach 23 million MT by the end of the Tenth Plan. Urea and phosphate are largely produced domestically, while potash is imported. Chemical fertilisers are subject to 16 per cent excise duty. However, fertiliser companies receive a subsidy or concessions to promote the use of fertilisers by farmers. Gradual reduction of the subsidy or concessions is under consideration for fiscal and other reasons. A reduction of the subsidy would be more beneficial for the environment, rather than levying an ecotax on chemical fertilisers.

The production of biofertilisers has stagnated at around 700 MT (Fertiliser Association of India, 2003). Animal or vegetable fertilisers are already exempt from excise duty. Promotion of biofertilisers would require research and development to improve their quality and improve their shelf life. This has been recommended by the working group on fertilisers for the Tenth Plan. Government should formulate and implement a fairly comprehensive policy, (a) to improve the quality of biofertilisers by providing capital subsidy for investment in R&D to identify more suitable strains, and to develop better production technology and quality control methods, and (b) to encourage the states to undertake extension work in respect of use of biofertiliser. The medium and long term aim should be to raise the use of biofertilisers to the scale implied in the integrated plant nutrient management system.

9

Lead Acid Batteries

Lead is used in many applications such as batteries, cable sheathings, and pipes. However, lead acid storage batteries are the largest tonnage consumers of lead accounting for about 70-75 per cent of the lead used in India. Lead demand is met from two sources: primary and secondary sources. Primary sources are lead production from lead ore and import of lead, whereas secondary sources consist of lead smelters, which recover lead by recycling of lead scrap/waste, primarily used lead batteries. The production of lead in India from primary sources accounts for nearly two-thirds of the total lead production in the country. According to Rajagopalan (2002), in the world as a whole, lead from secondary sources accounts for nearly 60 per cent of the total production of lead. The lower share of secondary lead in India could partly be explained by the presence of a large number of small unregistered lead smelters characterised by poor lead recovery rate (30-40 per cent).

Indian Battery Industry

According to India Lead Zinc Development Association (ILZDA), 7-8 battery manufacturers in the organised sector cater to about 45 per cent of the lead acid battery market in the country. Medium and small-scale manufacturers numbering around 250 and as many as 4,000 battery assemblers/re-conditioners cater to the rest of the demand. Total sales, market shares and imports and exports of the large companies are given in Tables 9.1-9.2 below. It may be seen from Table 9.1 that the total sales of these companies have increased by over 58 per cent during 1996-97–2001-2002, that is, an average increase of 9.7 per cent per annum. Among these companies Exide Industries held 50-60 per cent of the market share.

Table 9.1

Company-wise Trends in Sales: 1996-97 to 2001-02

(Rs. Crore)

	1996-97	1997-98	1998-99	1999-00	2000-01	2001-02
Exide Industries	501.91	556.12	812.75	934.66	948.58	968.48
Amara Raja Batteries	101.01	144.38	186.55	132.08	155.01	188.22
H.B.L. Nife Power Systems	15.34	21.55	34.44	85.15	126.78	145.32
Amco Batteries	84.54	70.23	70.23	63.32	63.14	63.14
Tidor India	8.67	17.77	28.33	37.24	43.18	44.91
Tractors & Farm Equipment	22.74	17.12	23.41	26.34	28.40	29.40
High Energy Batteries (India)	13.15	18.05	19.15	20.11	11.94	14.66
Amco Power Systems				3.25	7.40	8.50
Willard Storage Battery	17.76	1.22	0.09			
Standard Batteries	128.24	110.71				
Nicco Corpn.		13.37	14.90	7.93		
Kirloskar Batteries		23.70	23.70			
Herbertsons	30.14					
Eddy Current Controls (India)	0.23	0.12	0.05			
Total	**923.73**	**994.34**	**1213.50**	**1310.08**	**1384.43**	**1462.63**

Note: Willard Storage Battery was declared as sick unit by BIFR in 1999. Nicco Batteries was merged with Nicco Corp. in 1995-96. Kirloskar Electric acquired the energy products division (UB-MEC Batteries) of Herberssone during 1995-96 and named it as Kirloskar Batteries. Exide Industries acquired the entire business of batteries of Standard Batteries with effect from February, 1988/ H.B.L. Ltd. and Pinaki Technologies were merged with Sab Nife Power Systems effective from April, 1999 and the name of the company has been changed to HBL Nife Power Systems Ltd.

Source: India Lead Zinc Development Association.

The automobile industry and certain other industrial users are the main consumers of lead acid batteries. Typically, the distribution between automobile and other industrial applications is 60:40 in terms of the value of batteries and lead consumption (IBMA, 2001). In these sectors the demand for batteries comes from the following:

(i) Automobiles: Cars/jeeps and SUVs, buses, trucks and tractors, two and three-wheelers

(ii) Other industrial applications: Fork lift trucks, railways, power stations, telecommunications, UPS/computer applications, defence and submarine applications.

Table 9.2

Trends in Market Share: 1996-97 to 2001-02

(Per Cent)

	1996-97	*1997-98*	*1998-99*	*1999-00*	*2000-01*	*2001-02*
Exide Industries	51.23	51.74	62.06	65.75	60.95	58.04
Amara Raja Batteries	10.31	13.43	14.24	9.29	9.96	11.28
H.B.L. Nife Power Systems	1.57	2.00	2.63	5.99	8.15	8.71
Amco Batteries	8.63	6.53	5.36	4.45	4.06	3.78
Tidor India	0.89	1.65	2.16	2.62	2.77	2.69
Tractors & Farm Equipment	2.32	1.59	1.79	1.85	1.82	1.76
High Energy Batteries (India)	1.34	1.68	1.46	1.42	0.77	0.88
Amco Power Systems				0.23	0.47	0.51
Willard Storage Battery	1.81	0.11	0.01			
Standard Batteries	13.09	10.30				
Nicco Corp.		1.34	1.14	0.56		
Kirloskar Batteries		2.21	1.81			
Herbertsons	3.08					
Eddy Current Controls (India)	0.02	0.01				
Total for the above companies	94.29	92.51	92.67	92.16	88.96	87.65
Total for the sample 16 Cos.	94.29	92.51	92.76	92.29	89.07	87.65
Import	5.07	6.96	6.84	7.14	10.04	10.11
Herfindahl Index of Concentration	0.300	0.302	0.410	0.448	0.391	0.360

Source: India Lead Zinc Development Association.

According to an estimate (IBMA, 2001), in 2002-03, the demand for batteries from the automobile sector was 80.9 lakhs. Of this, 13.4 lakh batteries were supplied to the original equipment manufacturers (OEM) and the remaining constituted the replacement demand. The industrial demand for batteries was 32.36 lakh. Ten per cent of the total industrial demand and 6 per cent of the total demand from automobiles was met by imports.

Indian Lead Sector

While production of lead in India has declined by about 10 per cent during 1992-93–2000-01, the consumption of lead has more than doubled during the same period (Rajagopalan, 2002). It is important to note that during 1997–2000, lead consumption in India grew at the rate of 10.5 per cent against 2.5 per cent world average growth rate. Rajagopalan (2002), assuming a 6 per cent annual growth in demand for lead, has estimated the demand for lead at 3,16,000 MT in the year 2016-17, and the supply-demand gap to be around 1,64,200 MT. It may be noted that these projections do not take into account the demand for lead in the unorganised sector engaged in assembling/reconditioning of batteries and also the supply of lead from the unorganised sector consisting of backyard smelters and small smelters. A brief description of these is given below.

Battery Re-conditioners

Battery re-conditioners test 'used' batteries to ascertain whether the battery can be reconditioned by just recharging it or whether one or more of the battery plates require replacement. Batteries with defective plates, terminals, etc. require replacement of these parts. Re-conditioners source their requirements of lead and battery plates from the smelters in the unorganised market.

Backyard Smelters

They typically break open used batteries to segregate pieces of plastic, battery plates, etc. The battery plates complete with the lead oxide and sulphate pastes still embedded in the battery plate grids are melted in large open kilns or cast iron pots. The metallic lead grids are melted and the metal is tapped and cast in moulds to produce unrefined lead ingots. The rubber and polypropylene cases are sold to plastic recyclers for eventual resale to the battery manufacturers.

Small Smelters

The small smelters smelt the plates and many of them recover up to 90 per cent of the available lead. A number of units have the capacities to produce more than 1000 kg of lead per day. These units operate without observing the approved pollution control measures. Organised sector battery manufacturers are reported to purchase large quantities of lead produced by these units.[1] This points towards the need to mandate even the organised sector battery manufacturers to source their lead from the approved recyclers.

1. Minutes of the Meeting of the Core Group on Batteries, February 1, 2002.

Table 9.4

Lead Production

Metric Tonnes (MT)

	1998-99	1999-2000	2000-01	2001-02
Hindustan Zinc Ltd. (HZL)	39010	35120	15238	37860
Indian Lead Ltd. (ILL)	8902	7347	-	-

Source: India Lead Zinc Development Association, Minerals and Metals Review (various issues).

Primary Lead Production

According to the information in *Minerals and Metals Review*, the Indian Lead Limited (ILL)—one of the two primary lead producing companies in India—ceased production of lead since 2000-01. As a result, the primary lead production in India was only 37,860 MT in 2001-02.

Table 9.5

Lead Demand and Supply with Existing and Planned Capacities—10th Five Year Plan

Year	Demand (Tonnes) with 6% Growth	Supply (MT)				GAP (MT)
		HZL	ILL	SEC. LEAD	Total	
2002-03	137700	38700	21600	45000	105300	-32400
2003-04	146000	70200	21600	45000	136800	-9200
2004-05	154700	70200	21600	50000	141800	-12900
2005-06	164000	70200	21600	50000	141800	-22200

Source: Tenth Five Year Plan Draft Report.

Tenth Five Year Plan Estimates

According to the Tenth Five Year Plan estimates, the demand for lead—both primary and secondary—would exceed the supply by 9,200 tonnes in 2003-04. The gap between demand and supply is estimated to increase to 22,200 MT in 2005-06.

IBMA Estimates

For the year 2002-03, the total demand for lead was estimated at 1.92 lakh MT (IBMA, 2001). Over 80 per cent of this was for use in battery manufacture. The lead supply was estimated as follows:

(i)	Indigenous Producers	MT
	a. HZL	35,000
	b. Others	20,000
(ii)	Imports (including alloys)	65,000
(iii)	Secondary lead smelters	71,000
	Total	**1,90,000**

It was estimated that about 50 per cent of (iii) above would be from smelters that have MoEF clearance (full/partial clearance) and the remaining half would be from smelters operating without the approval from the MoEF.

Secondary Lead Sector

There are only three to four secondary lead processing units in the organised sector. There are a large number of tiny units in clusters at 5-6 locations in India, numbering around 250-300, each having a smelting capacity of 500-1000 kg, approximately, per day.

According to India Battery Manufacturers' Association (IBMA), before the Batteries Legislation 2001, only 11.5 per cent scrap reached the smelters in the organised sector. The remaining 88.5 per cent landed with traders who channelled it to smelters in the unorganised sector or gave to smelters in the organised sector depending on the price. As such, the availability of scrap is represented by the size of the replacement market. The replacement market in India is divided into two broad segments—trade and non-trade. The non-trade segment represents the government departments/undertakings and institutional buyers. According to IBMA, the trade segment accounts for 91 per cent of the replacement market. The unorganised battery sector sources the battery scrap from both consumers and dealers. A summary of the current replacement market for the batteries is presented in Table 9.6 and a flow chart representing the current replacement market is presented in Figure 9.1.

According to the Planning Commission, during the Ninth Plan period, lead production in the secondary sector declined on two accounts. First, due to the ban imposed on the import of lead scrap under the Basel Convention. This resulted in decline in secondary lead production of Indian Lead Ltd. to about 50 per cent of its annual capacity. Second, lead production declined due to the closure of the

Vizag lead refinery plant by HZL primarily due to environmental reasons. Since India has poor primary lead resources and no viable lead deposit is expected to come up for commercial exploitation, there is a need to increase the capacity of the secondary lead sector.

Table 9.6

Battery Replacement Market—Some Estimates

(Per cent)

	Trade	*Non-trade*	*Total*
Organised Sector (Large and Medium)	26	5	31
Imports	6	-	6
Small-scale Sector (new batteries)	36	4	40
Re-builders	23	-	23
Total	**91**	**9**	**100**

Health Effects of Lead

Lead is a toxic metal. If absorbed into the body in large amounts, either through a single high exposure incident or through long term chronic exposure to lower amounts, it can result in severe adverse health effects (Table 9.7). The principal health effects of exposure to lead include anaemia, harmful impact on the nervous system, high blood pressure, damage to the kidney, and adverse effects on male and female reproductive systems. There is some evidence that lead and lead compounds are carcinogenic.

Table 9.7

Potential Health Impacts of Exposure to Lead

Anaemia	Central nervous system
Peripheral nervous system	Renal effects
Blood pressure	Reproduction
Carcinogenicity	Immune system

In order to protect people living or working in the immediate vicinity of plants handling lead or who are exposed to air-borne lead from other sources, for example lead in fuel, ambient air quality

Figure 9.1

Flow of New and Scrap Batteries in India

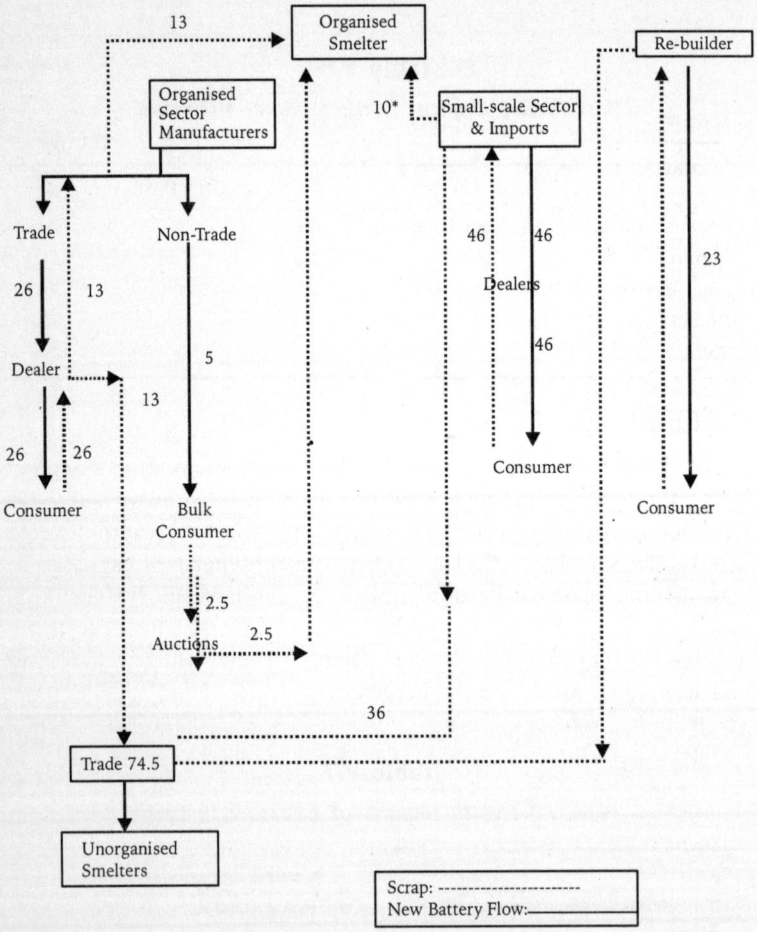

Note: * 10 = 4 (small scale non-trade sector) plus 6 (Imports). Figures are in percentages.

Source: *IBMA*.

standards have been set in many countries, including India (Table 9.8). Water also has the potential to constitute a significant source of lead intake. The drinking water standard in India is currently 100 μg/l.

Table 9.8

Ambient Air Quality Standards for Lead

Max. Lead Level ($\mu g/m^3$)	*Country*
4.0	South Africa
3.0	USSR
2.0	Belgium, France, Germany, Netherlands, Spain, United Kingdom
1.5	India (24 hourly average for industrial areas)
1.5	Australia

Source: International Lead and Zinc Study Group on "Environmental and Health Control on Lead" (year not given).

Table 9.9

Dealers' Incentive

	If He Gave to Company (Rs.)	*If He Gave to Trader (Rs.)*
Dealer Buys Back ScrapFrom Customer at	106	106
Price He Gets From Company	135	160-190
Margin on Scrap	29	54-84
Income Tax @ 30% (say)	9	-
Net margin	20	54-84

Source: *IBMA*.

Why Dealers Sell Battery Scrap to Traders and not Battery Manufacturers

The higher profit margin in the table above explains dealers' incentive to sell battery scrap to the trader rather than the manufacturer.

For the organised sector battery manufacturers/smelters, only the lead in the scrap batteries has some value, whereas for the re-builders every part of the scrap battery has a value. This is because the re-conditioners/assemblers are able to make use of the hard rubber container, negative plates, separators, etc. in addition to the lead content of batteries and thus gain an advantage *vis-à-vis* the organised sector. The following table shows the cost advantages of the informal sector.

Table 9.10

Cost Structure in Different Sectors

(in Rs.)

	Organised Sector	*Small Scale*	*Re-builder/ Re-conditioner*
Consumer Pays	100	80	60
Sales Tax	(11)	(9)	-
Distribution Costs (dealer margins, etc.)	(20)	(10)	-
Excise duty	(10)	(5) (concessional)	-
Realisation of Manufacturer	59	56	60

Source: *IBMA*.

Thus a re-builder can afford to sell the product 40 per cent cheaper and still realise the same amount as the organised sector.

Salient Features of Batteries (Management and Handling) Rules, 2001

In order to regulate collection of used lead acid batteries and channelise these batteries to authorised recycling facilities, the Batteries Rules, 2001 have been notified. The salient features of the rules are as follows:

(i) Consumers should return the used batteries to authorised traders/manufacturers/collection centres.

(ii) Manufacturers/assemblers/re-conditioners/importers shall be responsible for collection of batteries and transport to registered recyclers.

(iii) Auction of used batteries shall be only in favour of registered recyclers.

(iv) Dealers shall also be responsible for collection of used batteries.

(v) Batteries have been categorised by type of use.

(vi) The prescribed authority for ensuring compliance of the provisions of these rules shall be the State Pollution Control Boards (SPCBs). They shall file an annual compliance status report to the Central Pollution Control Board (CPCB) by 30th April of every year.

(vii) The CPCB shall compile and publish the data received every year from the SPCBs. It shall review the compliance of the rules periodically to improve the collection and recycling of used lead batteries and apprise the MoEF.

(viii) The MoEF or an agency designated by it shall develop a system for computerised tracking of (a) distribution, sale of lead batteries, and transport and reprocessing of used batteries; (b) sale of re-processed lead by registered recyclers, and sale of lead from all domestic producers or importers.

In addition to Batteries Rules 2001, a scheme for registration of re-processing of used lead batteries is in operation since 1999. Re-processing units are required to follow the prescribed code of practice for environmentally sound management (ESM) of lead batteries and possess proper facilities for disposal of water. The units are also required to register with the MoEF.

The Main Issues at Hand

There are three main issues to be addressed. First, checking the diversion of used batteries from the dealers to the unorganised sector (the main cause for the diversion of scrap batteries to the unorganised sector is a ready market for the scrap batteries at attractive prices). Second, identifying measures to check the demand for re-conditioned batteries, which according to the available estimates takes away 23 per cent of the total scrap generated. This sector also sources its lead from the unorganised sector thus contributing to the growth of this sector. Third, improving the competitiveness (through economic measures) of the smelters in the organised sector *viv-à-vis* those in unorganised sector who do not pay taxes and thus can offer relatively higher price for the scrap batteries

In order to address these issues, actions would be required on the following lines:

(i) Design and implement penal measures to enforce return of batteries by dealers to authorised smelters/battery manufacturers.

(ii) Design and implement penal measures to eradicate unauthorised smelters and unauthorised re-builders/re-conditioners.

(iii) Battery manufacturers should be required to penalise/incentivise dealers to sell scrap batteries only to authorised smelters/manufacturers.

(iv) Measures to penalise battery manufacturers for buying/sourcing lead from unauthorised smelters/producers of lead.

(v) Measures to integrate unauthorised re-builders/re-conditioners and smelters into the mainstream.

(vi) Design and implement economic measures to improve competitiveness of the organised smelters *vis-à-vis* those in the unorganised sector.

Measures to Control Environmental Pollution from Batteries: International Experience

Sweden

Sweden was the first country to introduce a Battery Return Scheme called Returbatt. Under the scheme, the government and Returbatt agree on a unit levy and annual support level. A levy of SKr 32 per battery heavier than 3 kg is collected from the battery manufacturers. The levy determination takes into consideration the distance between the only secondary lead smelter in Sweden and points of sale of more than 1200 km.

British Columbia

The scheme is based on 'an environmental levy' on all new lead acid batteries. The Government determines the level of the levy and the incentives for transport are determined on a monthly basis by a government-sponsored body. The existing collection infrastructure is utilised. The transportation incentive is available only to (authorised eco-friendly) processors of used batteries.

Italy

A non-profit organisation, COBAT, whose members come from the battery industry, secondary lead smelters, garages, merchants and government was set up. COBAT determines the levy on all new batteries. Licenced scrap dealers and smelters apply for support funds. The chief functions of the COBAT are: (i) to improve collection and recycling of batteries, (ii) to collect levy on new battery sales, and (iii) to educate the public on the need for ecologically sound disposal of spent batteries. The main difference from the Swedish system is that there is no obligation on the battery retailers to accept spent batteries.

To finance its activities, the consortium has two sources of income: (i) receipts from sale of scrap batteries to the recycling plants, and (ii) a levy on the sale of new batteries. The levy on battery sales is set by the government and reviewed periodically to ensure an adequate level of funding for the proper functioning of the consortium.

United States

The Battery Council International (BCI) is actively promoting adoption of a proposed model for battery recycling legislation by all the individual states. The main elements of the BCI model comprise (i) prohibition of land disposal of batteries, and (ii) acceptance of spent batteries by battery retailers. In order to make disposal easy for the owner of a spent battery; the model requires all retailers of new batteries to accept old ones in whatever numbers offered.

Recommendations

Recycling of scrap lead, primarily used batteries, and using certified methods is necessary for environmental protection. In order to increase the size of this activity in the organised sector, the supply of scrap batteries to this sector needs to be increased. This implies that unauthorised recycling of scrap batteries, and re-building/re-conditioning of batteries needs to be checked. The enforcement of current legislation, which focuses only on the organised sector manufacturers to organise for the return of used batteries to the authorised smelters can be expected to have only a limited impact in the shortrun or even in the medium term, because the unorganised market will continue to source scrap batteries from the market it caters to (assuming that no diversion of scrap batteries from the dealers to the unorganised re-cyclers and re-builders takes place). To meet the demand for lead from this sector, there will be illegal smelting activity.

It is important to note, however, that the existence of smelters in the unorganised sector is not only due to the battery re-builders and re-conditioners in the unorganised sector, but also due to the battery manufacturers in the organised sector who are reported to source their lead from smelters in the unorganised sector.[2] Thus, both the dealers and the battery manufacturers in the formal sector have a crucial role to play in checking the proliferation of battery re-building and smelting activity in the unorganised sector.

Strict punitive measures should be taken against unauthorised battery manufacturers, and re-cyclers/re-conditioners. Such measures should be made a part of the Battery Rules. Progressively increasing targets should be fixed for the enforcement authorities, which in this case are SPCBs. Punitive measures for the dealers and manufacturers should they fail to meet the targets for collection of scrap batteries should be explicitly outlined in the Battery Rules. It should be the responsibility of the enforcement agencies to take appropriate and timely action against the violators. This would lend credibility to the spirit and intent of the Battery Legislation such that it acts as a real deterrent.

To motivate the enforcement agencies for achieving the targets, it is important that the progress in respect of collection of scrap batteries, and action taken against the unauthorised activities of battery manufacturing, re-building, lead recycling are closely and systematically monitored and recorded. This should then be reviewed periodically to fix the accountability and identify additional measures required to achieve the targets.

From an environmental point of view, scrap battery recycling by the licenced smelters needs to be encouraged. India has a large informal sector comprising battery re-conditioners/re-builders and recyclers. Since they do not pay taxes and have insignificant costs for pollution abatement and occupational safety *vis-à-vis* licenced smelters, they can offer higher prices for battery scrap than the formal sector would afford. According to a study (Chaudhary and

2. Minutes of the Meeting of the Core Group on Batteries, February 1, 2002.

Hoffman, 2001) scrap battery prices in India are 2-3 times higher than the scrap prices in the international market. The informal sector operations thus adversely affect the smelting activity in the formal sector by making it economically unviable.

Therefore, strict enforcement of the regulatory measures as outlined above needs to be complemented with economic measures to improve the competitiveness of the licenced recyclers *vis-à-vis* informal sector smelters. The following measures are proposed:

i) reduction in excise duty on production of secondary lead by the organised smelters.

ii) levy of an environment cess on the sale of scrap batteries in auctions by the bulk consumers. The environment cess should be allowed to be set off against the levies on production of secondary lead. This could be expected to check the participation of the unorganised smelters/traders in auctions.

Experience has shown that the regulatory measures can have only a limited impact in discouraging the production of items that have a substantial demand in the market. In this context, it is important to understand why India has such a proliferation of battery re-conditioners/re-builders that are estimated to control 23 per cent of the battery replacement market.

There is a sizeable market in India for re-conditioned batteries because of their substantially low price *vis-à-vis* new batteries. Encouragement to low price battery line with relatively higher product life (12-16 months) would provide an alternative to the consumers of re-conditioned batteries. Furthermore, the development of batteries with a longer life (6-8 years) than that of currently produced batteries (2-3 years), may also progressively discourage consumers from buying reconditioned batteries. This would also reduce the number of batteries in the recycle loop. The markets in low price battery line and 'long-life' lead acid batteries can therefore be the driving forces to displace re-conditioned batteries.

At the same time, government should take measures to encourage the integration of the reconditioning sector into the mainstream recycling sector or used battery collection sector in which they have lot of experience.

A comprehensive strategy to integrate the informal sector into the mainstream would require better understanding of the informal sector including the size of the informal sector, and the number and size of the industrial units/enterprises, technology used, sources of feed stock, output and inter-relationship between these groups (battery re-conditioners, backyard smelters, and small lead smelters). This information is however not available. In view of the limited information available in respect of the informal sector, the following broad suggestions are made to create and obtain more information.

Define a role for the re-builder/re-conditioner and small smelter. Identify pockets that have sizeable re-conditioned battery market. As the size of the reconditioned battery market shrinks as a result of enforcement of both the regulatory measures and the demand side measures outlined above, owners of these units can be encouraged to join hands together with the small smelters for setting up secondary lead consortium and upgrade the existing process such that their recycling facilities conform to the certification requirements of the MoEF. These units will need to be helped by the regulators with affordable technologies.

Battery manufacturers should be required to incentivise the return of scrap batteries by the dealers to the approved smelters. Manufacturers should operationalise this by making the scrap return to authorised smelters/manufacturers a part of the terms and conditions for the dealers and also by linking the dealers' margin to the return of scrap batteries.

Revenue Implications

A 50 Per Cent Reduction in Excise Duty

In the absence of information on excise revenue collected from production of secondary lead, it has been estimated based on the information from ILZDA, IBMA, and other sources as follows:

Demand for lead (for batteries)	: 1,75, 000 MT
Primary lead + imports	: 1,00,000 MT
Secondary lead	: 75,000 MT

Based on the available information on the flow of scrap batteries to the smelters in the formal sector about 35,000 MT is produced in the formal sector and the remaining is supplied by the unorganised sector.

Lead production attracts excise duty at the rate of 16 per cent. Taking the value of lead at Rs. 50 per kg, estimated excise revenue from lead works out to Rs. 28 crores (concessions in duty currently available are not taken into account). A 50 per cent reduction in excise duty rate would imply a revenue loss of Rs. 14 crores. However, as the competitiveness of the lead smelters improves, size of this activity is expected to increase resulting in increase in revenue base.

Also, contributions from the proposed environmental cess on scrap would reduce the net impact further.

10

Plastics

Introduction

Plastics are used in a wide variety of applications due to their durability and resistance to weathering. In India, plastics are reused and recycled much more widely than in many other parts of the world. However, some plastic products like carry bags, beverage containers and thin sheets are causing significant solid waste problems, particularly in urban areas. There is also concern that recycling may not be environmentally safe. The Government of India has formulated rules on thickness of carry bags and has issued guidelines for the recycling of plastics. In this chapter, we assess the desirability and feasibility of an ecotax and examine the concept of extended producer responsibility for certain plastic products.

Background

Plastics are synthetic polymers composed of large molecules formed by joining small molecules (monomers). They have the ability to be shaped/moulded to form articles of practical value. Plastics have virtually permeated every sector—agriculture, building and construction, packaging, education, transport, etc. Globally, the consumption of plastics has been growing at 6 per cent per annum during the last decade. In India, the growth rate has been much higher (12-15 per cent per annum) with consumption trebling every decade. Consumption has grown from 15,000 tonnes per annum in 1961 to 4 million tonnes per annum in 2001. The elasticity with respect to GDP was of the order of 2.2 for the period 1993-2001. Thus, plastic use has been growing at more than double the rate of growth of output of the Indian economy (ICPE, 2003).

Plastics are mostly derived from crude oil and natural gas. After refining, the building blocks like the olefins are converted into polymeric compounds such as polyethylene and polypropylene. These polymers are first produced as powders and granules, and the application of heat and pressure results in the final product. Plastics are generally classified as thermoplastics or thermosetting resins. Thermoplastics which constitute about 85 per cent of plastics can be melted and reshaped. Thermosetting resins on the other hand cannot be recycled because they undergo an irreversible chemical reaction.

The four major types of thermoplastics produced in India are polyethylene, polypropylene, polyvinyl chloride and polyethylene terephthalate. Table 10.1 contains the production data for these plastics in thousand tonnes per year. It is to be noted that about half of the production (52 per cent) is used for packaging purposes.

Table 10.1

Production of Polymers

(in Thousand Tonnes Per Year)

Polymer	*1995-96*	*2001-02*	*2006-07 (projected)*
Polyethylene (PE)	823	1835	3267
Polypropylene (PP)	340	885	1790
Polyvinyl Chloride (PVC)	489	867	1287
Polyethylene Terephthalate (PET)	34	140	289
Others	203	647	1415
Total	**1889**	**4374**	**8054**
Packaging Use	976	2272	4037
Per Cent Use in Packaging	(52%)	(52%)	(50%)

Source: National Plastic Waste Management Task Force Report (1997).

The per capita production of plastics in India which is about 3.6 kg–4.0 kg per year, is expected to reach 8 kg in 2010. This level is much lower than the world average of 21.5 kg per capita. Production of plastics from virgin material may be lower in India since 60 per cent of used plastics are recycled.

The major raw material manufacturers of plastics in India are Reliance, GAIL, IPCL, BASF, Bayer, and Finolex Industries. Plastic processing plants are comparatively smaller. There are approximately 22,000 units across the country in the small/medium sectors.

Recycling

More than 1 million metric tonnes per year of plastics are recycled. It is estimated that there are 2300 recycling units generating a value added of about Rs.1600 crore and employing about three lakh people. Recycling is generally done using primitive technologies. The sorting of plastic scrap is done on the basis of colour, transparency, hardness and opacity. The sorted waste is sent to the granulators, where extrusion is done to produce granules. The granules are then processed in converters into finished products (Narayan, 2001).

While recycling saves on virgin materials and eases the problem of disposal, it also causes several environmental problems:

- The wastewater from the cleaning of the scrap has high pollution load.
- Toxic dyes and chemicals are used as additives for recycling which are bound to affect the health of the workers.
- Contaminants from the first use may remain in the plastic and leach out later.

Recycled plastic is cost competitive with virgin plastic under Indian conditions due to the low cost of recycling. The price difference is due to the low cost technology and the low wages paid in the informal sector. There is a growing demand in India for inexpensive low quality plastic products, which is met by recycled plastic. Recycling in India is really 'down cycling' in the sense that the recycled product is inferior to the original product.

Plastic Wastes

The non-recycled part of the waste stream ends up as litter and forms part of municipal solid waste (MSW). In most Indian cities, plastics consist of 1-2 per cent by weight of MSW. Dealing in plastic waste in general is lucrative compared to paper, glass, etc. since it commands a higher price from waste dealers. However, there is not much demand for carry bags due to their low weight, and hence these end up as street litter. Plastic wastes for which there is a market end up with recyclers, including wastes from hotels and catering industry and from air/rail travel.

Ragpickers form an important link in the waste chain. They collect the plastic wastes which have market value and sell them to the waste dealers. It is not worthwhile for the ragpickers to collect carry bags due to their low weight. Clean carry bags fetch between Rs. 5-10/kg but it takes 200-350 bags to make a kilogramme of plastic bags (Narayan, 2001). It is suggested that the plastics industry should consider giving incentives to ragpickers and NGOs for increased collection of plastic wastes such as carry bags and thin sheets from public places (CPCB, 1998). The incentive could match the amount paid by the recycler, namely Rs. 10/kg for collecting and/or segregating the plastic bags. The industry should provide this incentive as part of its extended producer responsibility (See page 182).

Uncollected plastic wastes end up in the landfill and remain there in perpetuity because they are not biodegradable. They are consumed by animals like cows leading to obstruction of the intestinal system. Or they tend to litter the country-side causing an aesthetic problem. Some states have chosen to ban carry bags. However, there has to be consistency in regulations among the states for the ban to be effective. The Ministry of Environment and Forests has specified a minimum thickness of 20 microns for bags made from virgin plastics and 25 microns in the case of bags from recycled material.

Biodegradability

Biodegradability is the intrinsic capacity of a material to be degraded by the action of micro-organisms. The waste product is used as a carbon source by micro-organisms and converted to innocuous products like carbon dioxide, biomass and water. The use of biodegradable plastics would be appropriate for manufacturing plastic products used to package food or in biomedical applications where recycling is not possible. However, biodegradable polymers may not be universal substitutes for the common resins used in many applications. Biodegradable plastics can be considered for applications such as garbage bags, food packaging, etc. where recycling or other disposal options are not possible.

Bioplastics as an Alternative

Biodegradable plastics from renewable sources (bio-based) are a promising alternative to plastic products which have a short life cycle or are impractical to recycle. Bioplastics are plastics made from building blocks of complex carbohydrates like cellulose and starch using micro-organisms. The main advantage of bioplastics is that they are biodegradable and can be used in many applications where the plastic becomes a waste product. Biodegradation can occur in one of two ways. One way is to synthesise a polymer which is vulnerable to break down by micro-organisms. The second method is to incorporate additives which can break the molecular chain. The second method has been used more extensively in India. The market for bioplastics was about 46,000 tonnes in 2001 and it is expected to grow to 96,000 tonnes by 2006-07. Bioplastics are more expensive than synthetic polymers because of the high R&D involved as well as the present low scale of production. A firm in North India produces 30–40 tonnes/year of bioplastic which is used to make carry bags, shopping bags, disposable containers, etc. The selling price is Rs. 80-100/kg, on top of which is the excise duty of 16 per cent. The legal framework for the utilisation of biodegradable material is unclear, since the distinction between biodegradability and compostability has not been resolved (Narayan, 2001).

One of the concerns about biodegradable polymers is that they may enter the waste stream and get mixed with recyclable plastics. Biodegradability and recyclability are very different characteristics which are mutually exclusive. Thus, users have to be educated on how to separate the waste streams. Biodegradability must be application specific and is not needed for all polymer applications. There must be test methods and standards that can be used. An international standard for degradable materials is now being developed, which is vital for bioplastics to be used.

Other Disposal Options

a. *Road making*: A promising alternative for disposal of plastic bags is to mix them in the bitumen used in laying of roads. Some demonstration projects in Karnataka and Tamil Nadu indicate the feasibility of using recycled polyethylene from grocery bags, which improves the performance of the bituminous mixes used for surfacing roads. Greater durability and longer life have been reported (Justo and Veeraraghavan, 2002; ICPE, 2003).

b. *Incineration:* Plastic wastes have high calorific value, but in the Indian context they are mixed with municipal solid wastes, which contain a lot of moisture and organic matter. Waste to energy incineration plants require sorting out the plastic from the municipal solid waste. There is also concern about the air pollutants that are emitted from incinerators as well as the toxicity of the fly ash which has to be safely disposed. Western countries that are using incineration have advanced flue gas cleaning technologies. In some countries incinerators have been closed down due to public opposition. All these aspects need to be carefully weighed before proceeding with such waste to energy technologies.

Government Initiatives

To address the public and media concern, the Government of India set up a National Plastic Waste Management Task Force in 1996, since there was no definite policy or legislation with regard to plastic waste. The objectives of the NPWM Task Force were the following:

- Formulate a strategy and prepare an action programme for managing plastic waste.
- Propose incentives and penalties to check the growth of plastic packaging waste.
- Prepare guidelines for the use of plastic packaging in the country.

The Task Force came up with a 14-point action programme. The Guidelines for Plastic Packaging and Guidelines for Recycling were formulated by the Ministry of Environment and Forests (MoEF) and the Bureau of Indian Standards. The guidelines emphasise source reduction in packaging and also in production; and call for an organised system of return, collection, and recovery of plastic waste. Source reduction essentially involves minimising the use of plastics in packaging by reducing the thickness (Ratra, 1997).

In 1999, the MoEF has also notified the Recycled Plastics Manufacture and Usage Rules under the Environment (Protection) Act with the following provisions (among others):

- The use of carry bags and containers made of recycled plastic for storing, carrying, dispensing or packaging of foodstuff is prohibited. Only virgin bags of 20 micron thickness of natural colour can be used.
- The minimum thickness of virgin carry bags should be 20 micron while recycled carry bags should be of 25 micron minimum thickness. Recycled carry bags can be used for non-food applications.

- Recycling of plastics should be carried out according to the Guidelines for Recycling.

The State Pollution Control Boards are responsible for enforcement of the rules within their states.

The Government of India launched the eco-labelling scheme known as Ecomark in 1991 for identification of environment friendly products. Under the Ecomark scheme the requirement for plastic products is that the material for packaging should be recyclable or biodegradable. The 1999 Rules specify that manufactures of recycled carry bags should code/mark as per the Bureau of Indian Standards (IS 14534 : 1998) for recycling plastic, and should also indicate the percentage of recycled material used.

A Committee on Plastic Waste Disposal was constituted in 2001 under the Chairmanship of Justice Ranganath Misra, former Chief Justice of the Supreme Court. The Committee made several recommendations in 2002 which include *inter alia:*

- Ban on manufacture of plastic carry bags of size less than 8" x 12". (The recycling rules were amended to include this provision in Section 4 of the June 2003 notification.)
- Manufacturers of carry bags or containers either from virgin plastic or recycled plastic should register with the State Pollution Control Board. (The rules were amended to include the provision in Section 6 of the June 2003 Notification.)
- Local governments should discourage the public from littering by imposing on-the-spot fines.
- Import of plastic wastes should be banned.
- Industry should be made responsible to retrieve empty packaging material and have a proper disposal system.
- PET bottle manufacturers should establish collection centres within six months failing which a deposit refund scheme at Re. 1.00 per bottle should be imposed.
- Collection centres and proper recycling facilities for plastic wastes should be established by the concerned industry in consultation with local bodies within one year.
- Local bodies should provide bins for segregated plastics and there should be mechanised handling system along with compacting.

Several states have taken initiatives with regard to plastic wastes:

- Himachal Pradesh has placed a ban on plastic littering and imposes penalties.
- Jammu and Kashmir has banned the use of plastic bags which have been replaced by alternative materials like jute and paper.
- Maharashtra has passed a state law upholding the central law prohibiting manufacture of bags less than 20 microns.
- Goa has announced a bar on recycled bags less than 20 microns and passed the Non-Biodegradable Garbage Disposal Act.

Extended Producer Responsibility (EPR)

EPR means that the primary responsibility for waste generated during the production process and after the product is discarded, is that of the producer/manufacturer of the product. Plastic producers in India have been attempting to shift the responsibility of plastic wastes to consumers and municipal authorities. The plastic waste problem in India cannot be solved unless producers take some responsibility for the ultimate disposal of their products. (Toxics Link, 2003).

One way this can be done is by 'deposit refund' systems for items like PET bottles. In India, there is a good deposit refund system for glass bottles used for beverages. In the case of PET bottles unlike glass bottles it may not be possible to refill the bottles. The used bottles will have to be sent to a recycling unit where the bottles are shredded and the fibres used for other purposes. If consumers pay a deposit of Re.1 per bottle, there would be an incentive not only for the consumer but also for ragpickers to collect and return the bottles. However, the manufacturers would have to design a network of collection centres for this purpose. Deposit refund systems on packaged drinks exist in Germany and in some states in the US Other plastic wastes like personal computers and telecom equipment also could be covered under EPR.

Ecotaxes on Plastic Bags

Excise duty of 16 per cent is levied on plastics and plastic products in India. The states would have to levy the ecotax on carry bags. Under VAT retailers will pay the tax. The 'retailer' will be the last registered dealer, who will collect and pay the tax. He may sell to consumers and unregistered dealers. Himachal Pradesh has used a policy of differential taxation. Paper bag manufacturers have been exempted from sales tax, whereas a 30 per cent sales tax has been imposed on plastic bags. It is estimated that the total cost of the bag including the sales tax works out to only 38 paise per bag. The tax has not been effective, since it is not large enough to affect the behaviour of the consumer (Narayan, 2001).

Several countries have imposed a tax on carry bags at the point of sale to the customer, and collected the proceeds from the retailers. Countries which have imposed ecotaxes on plastic bags include:

- Ireland- All outlets have been charging a levy of 15 cents since March 2002. Exclusions cover reusable shopping bags sold for 70 cents or more, bags used to contain fresh meat, etc., loose fruit and vegetables, and bags used for ice. The retailer is obliged to pass on the levy to the customer. Plastic bag use has gone down by 90 per cent since the introduction of the tax. The revenue is applied towards environmental projects.
- South Africa-Introduced a levy on manufactures of plastic bags, to be collected by retailers at the point of sale (as in Ireland). The levy would be 46 ¢ (for 24 litre bag), 31 ¢ (for 12 litre bag), and 25 ¢ (for 8 litre bag) and a ban on bags thinner than 30 microns. Out of the levy, 2 ¢ would go to a company that would promote recycling.
- Australia-One of the major retailers charged consumers an additional 10 cents for the use of a plastic bag. The reduction in use at various outlets of the retail chain was immediate varying from 40 per cent to 95 per cent.

Discussions with the Plastics Manufacturers Association

The Chennai Plastic Manufacturers Association (CPMA) made the following comments with regard to plastic wastes:

- They were not in favour of a levy on plastic bags at the point of sale since it would be difficult to recover the levy from thousands of retailers (this argument would not be valid when the states adopt VAT).

- They agreed that a deposit–refund system on PET bottles (similar to glass bottles) may be workable if a network of collection centres could be set up.
- They felt that the cost of biodegradable plastics was high, and acted as a deterrent to widespread use.
- Household segregation of non-biodegradable wastes is the critical factor in managing low value plastic wastes like carry bags.
- The existing MoEF regulation/guidelines are not implemented uniformly across the states. There is diversion of low quality plastics to the states where implementation is lax.

The comments of the CPMA were considered in formulating the following recommendations.

Recommendations

1. *Biodegradable Plastics*: The 16 per cent CENVAT on biodegradable plastics be removed to promote their production and use. If the cost is reduced, it will become more competitive with conventional plastics.
2. *Deposit Refused on PET bottles*: As per the Justice Ranganath Mishra Committee recommendations, a deposit of Re.1 per bottle should be levied on PET bottles at the time of sale which can be refunded when the bottle is returned. Manufacturers would have to set up a network of collection centres which will collect the bottles and send them for recycling.
3. *Incentive to Ragpickers*: Households can be encouraged by their respective municipal authorities to segregate their plastic wastes and hand over the low value wastes to the ragpickers. The plastic industry can provide a matching incentive amount (say Rs.10/-per kg of plastic bags) to the ragpickers in addition to the amount that the recycler would pay. This would be part of the extended producer responsibility of the industry with regard to the waste.
4. *Recycling*: Municipalities can also set up a central facility/ complex with assistance from the industry to recycle low value plastic wastes in an environmentally sound manner (CPCB,

1998). It is recommended that 50 per cent reduction in customs duty be given to recyclers who wish to import equipment and machinery for upgradation of recycling technology for a limited period of 10 years.

11

Summary

1. Environmental pollution is a negative externality. It arises when consumption or production decision of one individual effects consumption and production decisions of other individuals.
2. The externality problem can be solved by regulatory approach or economic approach or a mix of both approaches. Command and control (CAC) regulation gives little flexibility to polluters to find least cost methods of pollution prevention and control. It can be justified when the perceived harm from an activity is very high as in production/use of highly hazardous materials or when the assimilative capacity of a region has reached a threshold level or when an irreversible ecological damage would result. The precautionary principle justifies CAC regulations.
3. Economic Instrument (EI) is any instrument that gives a wide range of options to polluters to achieve specified levels of environmental quality at least cost. A tax on polluting output equal to its marginal social damage signals producers and consumers about the social damage of the activity. Creation of markets for pollutants, eco-rating and public disclosures on the extent of environmental compliance can create market/ community pressures for compliance and also reduce transaction costs in finding solutions.
4. The pollution control regime prevailing in most countries belong to any one of these or a mix of the three types: (a) standards and regulation, (b) standards and charges, and (c) standards and permits.
5. There is consensus among economists that CAC regulation does not minimise aggregate compliance costs. Its enforcement is ineffective when penalties for non-compliance with the regulations are unrelated to the compliance costs.

6. Many international agencies have argued the case for use of EIs. The OECD accepted the 'polluter pays' principle in 1972. The World Commission on Environment and Development stressed the need for merging economics and environment in decision making. Agenda 21 of the Rio conference stressed the need for integration of environmental costs into economic activities and to include, wherever appropriate, the use of market principles in the framing of economic policies.
7. The domestic policy support for EIs is also growing. The Tax Reforms Committee (1992) recommended that excise taxes could be an useful instrument in dealing with externalities in the form of social costs. The Supreme Court has accepted the 'polluter pays' principle as a part of the basic environmental law of the land. The MoEF Policy Statement for Abatement of Pollution, 1992 recommends 'new approaches for considering market choices to give industries and consumers clear signals about the cost of using environment and natural resources. It appointed a Task Force to evaluate MBIs for industrial pollution abatement in 1995. It organised an international workshop on EIs in cooperation with the World Bank and CII to appraise the industry about the experiences of developing countries on industrial pollution control. The MoEF Task Force on EIs has recommended the introduction of taxes on inputs and outputs of polluting industries.
8. There are no legal or institutional obstacles to the introduction of taxes on inputs and outputs of polluting industries. Introduction of pollution charges may require an amendment to the Environment Protection Act. Also, we need the capacity for measuring and monitoring emissions.
9. The advantages of the environmental taxes/charges are that they not only help in achieving improving environmental quality but also generate additional revenues which could be used for environmental improvement. For non-point pollution, particularly pollution occurring at the consumption stage, a tax on input/output is an ideal instrument for controlling pollution. For point sources, we have to devise taxes/charges with rebates/ refunds for units complying with the regulations.
10. International experience in the introduction of the taxes reveals mixed results. For successful implementation, we need

legal and institutional capacity, political support, and involvement of the stakeholders. Issues in the design of the taxes are evaluation of the trade-offs between environmental and other goals, choice of the tax base, consideration of existing distortions in prices and taxes, the basis for determination of the tax rates, phasing of the taxes, incentives for compliance and earmarking of funds for environmental protection.

11. This proposal covers coal, automobiles, phosphate based detergents, chlorine in paper and pulp and rayon industries, pesticides, fertilisers, lead acid batteries and plastics.

Coal

12. Coal is the most important energy source. About 70 per cent of the coal is consumed in the power sector. Indian coal contains 30-40 per cent ash and moisture content. Even though the environmental problems in using coal with high moisture content are well documented, there are not adequate incentives for coal washing. Based on estimates of cost of beneficiation of coal up to 30 per cent, this study recommends rates of ecocess on different grades of coal varying from Rs. 20 to Rs. 50 per tonne of coking coal, and Rs. 50 to Rs. 70 per tonne of non-coking coal. Ecocess as per cent of final price of coal would be at most 25.0 per cent for non-coking coal and 8 per cent for coking coal. A clean coal fund may be created to utilise the revenue from the ecocess for setting up infrastructure for coal washing, selective mining and R&D to identify activities by gainful utilisation of coal ash and safe storage and disposal of the residual ash.

Automobiles

13. Vehicular pollution is an important source of air pollution in most cities. It is a non-point source of pollution. Government has taken many measures to improve the quality of automobiles and auto fuel. The Expert Committee on Auto Fuel Policy has laid down a road map regarding vehicle norms and fuel norms. We recommend resource tax on passenger cars and two-wheelers based on their fuel economy. The tax rate as per cent of the ex-factory price of two-wheelers ranges

from 0.0 per cent to 3.0 per cent, and of passenger cars/jeeps from 0 per cent to 5 per cent. We considered an emission tax on diesel cars on the assumption that diesel vehicles are more polluting than petrol vehicles, but after an examination of the available scientific and empirical evidence, we could not reach the conclusion that the overall pollution load from a diesel car is greater than the pollution load from a similar petrol vehicle. Hence we defer the imposition of such a tax. We do recommend a tax on diesel cars to neutralise the price advantage in favour of diesel, if it is not possible to eliminate the price advantage of diesel. For in-use vehicles we recommend (i) an annual emission fee which may increase with the age of vehicle, (ii) introduction of emission warranty programme, and (iii) clean alternatives to be specified and certified for use by the regulator.

Chlorine

14. Chlorine used in pulp and paper and viscose rayon industries results in discharge of organochlorine compounds which are highly toxic. These mills could be encouraged to move towards chlorine substitutes like hydrogen peroxide if the rebate on the CENVAT of chlorine is withdrawn. However, the rebates can be continued for chlorine substitutes (even if they contain small quantities of chlorine) like chlorine dioxide or hypochlorite. The impact of the withdrawal of the VAT rebate on chlorine is likely to be small, since the cost of chlorine is only about 10 per cent of the cost of production. Since, switching over to chlorine substitutes will require substantial investments, accelerated depreciation at the rate of 50 per cent be provided in such machinery, which will lead to reduction in the use of chlorine.

Detergents

15. Phosphates have been used as 'builders' to soften the water in order to improve the cleansing action. But it contributes to an oversupply of nutrients to water bodies, and hence to the eutrophication of lakes and ponds. The three top brands of detergents are found to contain more than 20 per cent of sodium tri-poly-phosphate (STPP) by weight while the

preferred percentage is five or less. In a few countries phosphates in detergents are banned while in a few others phosphate use is restricted. There are alternatives to phosphates e.g. synthetic detergent zeolites. It is recommended that use of non-phosphate detergents like zeolite be encouraged by reducing the CENVAT on them from 16 per cent to 8 per cent. Non-rebate on VAT, on STPP may be considered to discourage the use of phosphate-based detergents. Compact detergent powders which save energy and materials may be promoted by decreasing the CENVAT on them to 8 per cent. Import of innovative pollution-free detergent technologies may be encouraged by reducing the import duties on such machines.

Pesticides

16. Spraying of chemical pesticides results in residues in drinking water, vegetables, milk, fish, etc. There is also occupational exposure to workers who apply the pesticides. At present, use of chemical pesticides is concentrated on a few crops like cotton, rice and pulses, and fruits and vegetables. There is a growing demand for organic farming and organic food. Integrated pest management is also being recommended. At present, the market for biopesticides is about one per cent of the pesticide market. *Neem* based pesticides dominate the biopesticides market in India. Organic fertililsers are exempt from excise duty, but there is no preferential treatment for organic pesticides. It is recommended that the CENVAT on biopesticides be decreased from 16 per cent to 8 per cent. This tax concession will make the relative price of organic pesticides cheaper, stimulate organic farming and increase export demand for organically grown agricultural products. An ecocess on chemical pesticides based on the level of toxicity is recommended.

Fertilisers

17. Decreasing the subsidy on chemical fertilisers will be more beneficial than imposing an ecotax as such. There is also a strong case for promoting the use of biofertilisers. Government should formulate and implement a fairly

comprehensive policy (a) to improve the quality of biofertilisers through capital subsidy for investment in R&D, and (b) to encourage the states to undertake extension work to popularise the use of biofertilisers. The medium and long-term aim should be to raise the use of biofertilisers to the scale implied in the Integrated Plant Nutrient Management System.

Lead Acid Batteries

18. Lead is a toxic material. There is some evidence that lead and lead components are carcinogenic. Recycling of scrap lead, primarily used batteries, and using certified methods is necessary for environmental protection. In order to improve the competitiveness of the licenced recyclers *vis-à-vis* informal sector smelters, we recommend (a) reduction in CENVAT on production of secondary lead by organised smelters, and (b) levy of an environmental cess on the sale of scrap batteries in auctions by the bulk consumers. This cess should be allowed to be set off against the levies on production of secondary lead. Battery manufacturers should incentivise the return of scrap batteries by the dealers to the approved smelters.

Plastics

19. Plastic products like carry bags, beverage containers and thin sheets are causing significant solid waste problems. There is a concern that recycling may not be environmentally safe. We recommend (a) removal of 16 per cent CENVAT on biodegradable plastics, (b) a deposit refund scheme at the rate of Re.1 on PET bottles at the time of sale, (c) an incentive scheme to ragpickers, and (d) 50 per cent rebate on customs duty on imported upgraded recycling technologies for use-in facilities set up by the recyclers.

References

Agnihotri, N.P. (1999). *Pesticide Safety Evaluation and Monitoring*, IARI, New Delhi.

Alam, G. (2000). "A Study of Biopesticides and Biofertilisers in Haryana", International Institute for Environment and Development, *Gatekeeper Series* No.93, UK.

Alberini, A. *et al.* (1998). "Fleet Turnover and Old Car Scrap Policies", *Discussion Paper* 98-32, RFF, Washington DC.

Anderson, R.C. and A. Carlin (1999). *Economic Savings from Using Economic Incentives for Industrial Pollution Control*, Environmental Law Institute Washington DC.

Ando, Amy *et al.* (1998). "The Economies of Vehicle Emission Repair", Paper presented at the *World Congress of Environmental Economists*, Venice, June.

Awasthi, U.S. (not dated). *"Current Status of the Fertiliser Industry in India–Policy Environment and Implication for the Future"*, mimeo, *IFFCO, New Delhi.*

Barret, A., J. Lawor and S. Scott (1997). *The Fiscal System and the Polluter Pays Principle – A Case Study of Ireland*, Ashgate.

Baumol, W.J. and W.E. Oates (1987). *The Theory of Environmental Policy*, 2nd edition, Cambridge, Cambridge University Press.

Bellomo, S.J. and S.D. Liff (1984). *Fundamentals of Air Quality for Highway Planning and Project Development*, US Department of Transportation, Washington DC.

Bhatnagar, P.R. and B.R. Sharma (2002). Groundwater Pollution through Agricultural Practices and Agro Industries in India, *IWMI-TATA Water Policy Research Program Annual Partners' Meet, Gujarat, India.*

Bohm, R.A. (1996). "Environmental Taxes–China's Bold Initiatives", *Environment* 40(10), September.

Brandon, C. and K. Homman (1995). *The Cost of Inaction: Valuing the Economy-wide Cost of Environmental Degradation in India*, Asia Environmental Division, The World Bank, Washington DC.

Bregha, F. and J. Moffet (1995). "The Tax for Fuel Conservation in Ontario", in R.Gale, S.Bang and A.Gillies (eds.), *Green Budget Reform*, Earthscan, London.

Capitaline (2002). *Capitaline 2000–An Electronic Corporate Database from Capital Market*, Mumbai, India.

Centre for Science and Environment (2000). "Green Rating Project: Indian Pulp and Paper Sector", *Down to Earth* 9 (12).

———. (2004). *Poison Vs Nutrition*, CSE, New Delhi.

Chaudhary and Hoffman (2001). "The Effectiveness of the Provision on Transfer of Technology to Developing Countries in the Basel Convention, the Case of Used Lead-Acid Batteries in India", Unpublished manuscript.

CIL (2001). *Coal India Limited Annual Report 2001*, New Delhi, India.

CMIE (2001). *Industry: Market Size and Shares*, Centre for Monitoring Indian Economy, Mumbai, India.

———. (2002). *Industry: Market Size and Shares*, Centre for Monitoring Indian Economy, Mumbai, India.

———. (2004). *Industry: Market Size and Shares*, Centre for Monitoring Indian Economy, Mumbai, India.

Coase, R.H. (1960). "The Problem of Social Cost", *Journal of Law and Economics*, 3: 1-44, University of Chicago.

CPCB (2000). "Clean Coal Initiatives", *Parivesh* Newsletter, Central Pollution Control Board, New Delhi.

Cropper, M.L., N.B. Simon, A. Alberini and P.K. Sharma (1997). "The Health Effects of Air Pollution in Delhi, India", *World Bank Policy Research Working Paper No. 1860* accessed from *http://www.worldbank.org/html/dec/publications/ workpapers/WPS1800series/ wps1860.pdf.*

Divan, S. and A. Rosencranz (2001). *Environmental Law and Policy in India*, Oxford University Press, New Delhi.

Down to Earth (2003). "Cover Story", September 15.

Driver, J. (2001). "The Arguments For and Against Using Phosphate as Detergent Builder", in R.K. Trivedi (ed.), *Managing Environmental Risks*, Tata McGraw-Hill Publishing Company.

Edwards, B. and David Santillo (1996). *The Stranger: The Chlorine Industry in India*, Greenpeace International, Amsterdam, Netherlands.

Eskeland, G. and S. Devarajan (1996). *Taxing Bads by Taxing Goods: Pollution Control with Presumptive Charges*, Washington DC, World Bank.

Fertiliser Association of India (2003). *Fertiliser Statistics*, FAI, New Delhi.

Francis, P., J. Klarer, and N. Petkova (1999). *Sourcebook on Environmental Funds in Economies in Transition: A Regional Overview and Surveys of Selected Environmental Funds in Central and Eastern Europe and the New Independent States*, Paris, OECD.

Ghosh, N. (2001). "Promoting Biofertilisers in Indian Agriculture", *mimeo*, Institute of Economic Growth, New Delhi.

Government of India (Ministry of Environment and Forests) (1992). *Policy Statement for Abatement of Pollution*, New Delhi.

———. (Ministry of Finance) (1992). *Report of the Tax Reforms Committee, An Interim Report* (Chaired by Raja J. Chelliah), New Delhi.

———. (Planning Commission), *Ninth Five Year Plan*, New Delhi.

———. (Ministry of Environment and Forests) (2001). *Battery (Management and Handling) Rules*.

———. (2002). *Expert Committee Report on Auto Fuel Policy*.

Harrington, W., *et al.* (1996). "Economic Incentive Policies under Uncertainty: The Case of Vehicle Emission Fees", *Discussion Paper 96-32*, RFF, Washington DC.

Harrington, W. and Virginia D. McConnell (1999). "Coase and Car Repair: Who should be Responsible for Emission of Vehicles in Use", *Discussion Paper* 99-22, RFF, Washington DC.

HERA (2002). "Human and Environmental Risk Assessment on Ingredients of European Household Cleaning Products–Zeolites", accessed from *http://www.heraproject.com.*

Hoerner, A. J. and B. Benoit (2001). *Environmental Tax Reform–the European Experience*, Centre for A Sustainable Economy, Washington DC.

Huber, R.M., J. Ruitenbeek, and R. S. Da Motta (1998). "Market Based Instruments for Environmental Policy Making in Latin America and the Caribbean: Lesson from Eleven Countries", *Discussion Paper* No. 381, Washington, World Bank.

India Battery Manufacturers Association (IBMA) (2001). "Issues in Battery Management and Handling Rules", Paper presented at the UNCTAD conference in Thailand, Bangkok.

ICRA (2003). *ICRA Industry Watch Series*, ICRA Information, Grading and Research Service, ICRA Ltd., New Delhi, *(www.icraindia.com)*.

India Lead Zinc Development Association (ILZDA). *Minerals and Metals Review, A News Round-up on Lead and Zinc*, various issues.

Indian Centre for Plastics in the Environment (ICPE) (2003). *Plastics for Environment and Sustainable Development*, New Delhi.

International Lead and Zinc Study Group on ''Environmental and Health Control on Lead'', Source of Table 8-quoted in *Global Trends in Monitoring and Controlling Lead Exposure*, V.R. Subramanian (2001), (Unpublished).

Justo, C.E.G. and A. Veeraraghavan (2002). "Utilisation of Waste Plastic Bags in Bituminous Mix", Unpublished Report, *www.ecologycenter.org/iptf/asia/ plasticroadreport.htm.*

Lovei, M. (1999). "Environmental Funds", Pollution Management in Focus Discussion Note No. 1., World Bank, Washington.

Mankiw, N.G. (1999). "Gas Tax Now!" *Fortune*, 24 May.

Markandya, A. (1993). "Environmental Taxation: A Review of OECD Country Experience and Prospects for Economics in Transition Development", *Discussion Paper* No.471, Harvard Institute for International Development, Harvard University.

Mehta, S., S. Mundle and U. Sankar (1993/1997). *Controlling Pollution: Incentives and Regulation*, Sage Publications, New Delhi.

Ministry of Chemicals and Fertilisers (not dated). *Draft Report of the Working Group on Fertilisers for the Tenth Plan*, Government of India, New Delhi.

Mukherjee, S. and S. Sud (2003). *A Comparative Study of the Impact of Emissions from Diesel and Petrol Fuelled Passenger Cars*, A Joint Study of Madras School of Economics and Delphi-TVS Diesel Systems Ltd.

Narayan Priya (2001). "Analysing Plastic Waste Management in India, Masters", Thesis, Lund University, Sweden.

National Institute of Public Finance and Policy (1997). *Report of the Task, Force to Evaluate Market-based Instruments for Industrial Pollution Abatement*, National Institute of Public Finance and Policy, New Delhi.

National Plastic Waste Management Task Force Report (1997). Accessed from *http://www.netpem.org/announcement/workshops/plastic/IndianPlasticManagementScenario.pdf.*

Norregaard John and V. Reppelin-Hill (2000). "Taxes and Tradable Permits as Instruments for Controlling Pollution: Theory and Practice", *IMF Working Paper* Wp/00/13.

O'Conner, D. (1998). "Applying Economic Instruments in Developing Countries: From Theory to Implementation", *Environment and Development Economics*, Vol.4.

Organisation for Economic Co-operation and Development (OECD) *(1992). Economic Instruments for Environmental Management in Developing Countries*, Paris.

———. (1994). *Managing the Environment: The Role of Economic Instruments*, Paris.

———. (1995). *Environmental Funds in Economies in Transition*, Paris.

———. Papers: 1996, 1997, OECD, Paris (*www.oecd.org*).

———. (2001). *Environmentally Related Taxes in OECD Countries–Issues and Strategies*, OECD, Paris.

———. (2002). *Environmentally Related Taxes in OECD Countries: Issues and Strategies*, Paris.

Olivecrona, C. (1995). "The Nitrogen Oxide Charge on Energy Production in Sweden", in Gale, R., S. Bang and A. Gillies (eds.), *Green Budget Reform*, Earthscan, London.

Opschoor, J.B. and H.B. Voss (1989). *Economic Instruments for Environmental Protection* OECD, Paris.

Pandey, R. (1988)."Fiscal Options for Vehicular Pollution Control in Delhi", *Economic and Political Weekly*, November 7, XXXIII, (45).

———. (2004). "Economic Policy Instruments for Controlling Vehicular Air Pollution", *Environment and Development Economics*, 9(1), February, pp.47-60.

Pandey and Bhardwaj, (2000). "Economic Policy Instruments for Controlling Vehicular Air Pollution", *mimeo*, National Institute of Public Finance and Policy, New Delhi.

Pigou, A.C. (1920). *Economics of Welfare*, Macmillan, London.

Rajagopalan, V. (2002). "Recent Indian Policy Initiatives in Lead Battery Scrap Management and this Impact on the Domestic Demand-Supply Gap of Lead", Paper presented at a conference in Bangkok.

Rao, E.V.S. and K. Puttanna (2000). "Nitrates, Agriculture and Environment", *Current Science*, 17(9) pp.1163-1168.

Ratra, O.P. (1997). "Plastics Waste Management: Systems and Policies", Proceedings of the National Seminar on *Recycling and Plastics Waste Management*, 24-26 September, J.S. Anand (ed.), CIPET, Chennai, India.

Ribeiro, M.T., K. Schlegelmilch and O. Gee (1999). "Environmental Taxes Seem to be Effective Instruments for the Environment", in *The Market and the Environment–The Effectiveness of Market-Based Policy Instruments for Environmental Protection*, pp.184-186, Thomas Sterner (ed.), Edward Edgar, UK and Massachusetts.

Sachdev, R. K. (1992). *Beneficiation of Power Grade Coals: Its Relevance to Future Coal Use in India, Cleaner Coal for Power*, URJA.32(1), India.

Sankar, U. (2002). "On the Design and Enforcement of Fiscal Instruments for Pollution Control in India", in M.Govinda Rao (ed.), *Development, Poverty and Public Policy*, Oxford University Press.

Sankar, U. and O.P. Mathur (1998). *Economic Instruments for Environment Sustainability*, National Institute of Public Finance and Policy and Madras School of Economics.

Schnutenhaus, J. (1995). *Tax Differentials for Catalytic Converters and Unleaded Petrol in Germany*, in R. Gale, S. Bang and A.Gillies (eds.), *Green Budget Reform*, Earthscan, London.

Sengupta, R., A. Lahiri and S. Mandal (2002). "Savings on Health Cost from Implementation of Euro Norms in India", *mimeo*, National Institute of Pubic Finance and Policy, New Delhi.

Shrotriya, G.C. (not dated). "Paradigm Shift in Fertiliser Sector in India", *mimeo* IFFCO, New Delhi.

Singer, J. G. (1981). *Combustion – Fossil Power Systems*, Combustion Engineering Inc., Windsor, Connecting, USA.

Singh S.P., and M.C. Bora (2000). *Alternative Technology for Clean Coal Recovery in Mining Policy Initiatives*, B. B. Dhar and N. N. Gautam, (eds.) MGMI, Delhi.

Smith, Stephen (1995). *Green Taxes and Charges: Policy and Practice in Britain and Germany*, Institute for Fiscal Studies, London, UK.

Society of Indian Automobile Manufacturers (SIAM) (2003). *Eco Tax Proposals – SIAM Views*, May 1.

TERI (2001). *Managing Fly Ash*, accessed from *http://teriin.org/energy/flyash*.

Tietenberg, T.H. (1985). *Marketable Emission Permits: Reforming Pollution Policy*, Resources for the Future, Washington DC.

Toxics Link (2003). *Breaking the Plastics Mould*, New Delhi.

United Nations, Conference on Environment and Development, (1992). *Agenda 21*, Rio de Janeiro.

Veerapan, V. (2001). "Zeolite–A Substitute for Phosphate Based Detergents", *in Managing Environmental Risks*, R.K. Trivedi (ed.), Tata McGraw-Hill Publishing Company.

Venkateshwarulu, S. and Anindya Sen (2002). "Fertiliser Industry in India: Moulded by Government Policies", *Economic and Political Weekly*, January.

Voice (2001). "How Green is Your Detergent", *www.consumer-voice.org*.

Weizsacker, E.U. von, (1992). *Ecological Tax Reform*, Zed Books.

White, M.J. (1982). "Pollution Taxes and Optimal Spatial Location", *Economica*, 499195: 297-311.

World Bank (1997). "A Planner's Guide for Selecting Clean-Coal Technologies for Power Plants", *World Bank Paper* No. 387, Washington DC, USA.

———. (2000), *Greening Industry, New Roles for Communities, Markets and Governments*, Oxford University Press, New York.

World Bank, Ministry of Environment and Forests and Confederation of Indian Industry (2002). *International Workshop on Economic Instruments for Industrial Pollution Prevention and Control in India, Proceedings*, New Delhi.

World Commission for Environment and Development (1987). *Our Common Future*, Oxford University Press, New York.

Xie, J., *et al.* (1998). "Fighting Urban Transport Air Pollution for Local and Global Good: the Case of Two-stroke Engine Three Wheelers in Delhi", Paper presented at Workshop on *Integrated Approach to Vehicular Pollution Control in Delhi*, 16-18 April, New Delhi, "Subgroup V: International Experiences in Vehicular Pollution Control."

Index

A

B

C

D

E

F

G

Q

R

S